INBREEDING, INCEST,
and the INCEST TABOO

Tables and Figures

Figures

INBREEDING, INCEST,
and the INCEST TABOO

Introduction

Arthur P. Wolf

The chapters of this volume are the fruits of a conference held at Stanford University, February 24–25, 2000. The conference was one of a series held to inaugurate the university's new Department of Anthropological Sciences. This was a happy—and not entirely fortuitous—conjunction of topic and occasion. The new department takes as its primary subject the relationship between biology and culture, and no point of contact between the two has generated more controversy than the relationship between inbreeding and the incest taboo.

This was not the first time scholars representing the biological and social sciences gathered at Stanford to discuss what is commonly called "the incest problem." In the spring of 1956, seven eminent researchers—David F. Aberle, Urie Bronfenbrenner, Eckhard H. Hess, Daniel R. Miller, David M. Schneider, and James N. Spuhler—organized a workshop at the Center for Advanced Study in the Behavioral Sciences "to consider the problem of the origins of the incest taboo."[1] Thus we have an appropriate and well-defined baseline to assess the development of the topic in the second half of the twentieth century. The changes are profound and will be readily apparent to anyone who compares Aberle et al.'s arguments with those developed by Patrick Bateson in Chapter 1 of this volume. I think these changes point to the direction anthropology is likely to take—and ought to take—in the new millennium.

In 1878, Mark Twain, traveling up the Rhine on a barge, came to a small town perched on "an instantaneous hill—a hill two hundred and fifty or three hundred feet high, and round as a bowl." It was Dilsberg, whose 700 inhabitants, Twain learned, were all "blood-kin to each other" and "have been blood-kin to each other for fifteen hundred years." The result, according to the captain of the barge, was that "for ages Dilsberg has been a thriving and diligent idiot factory." When, after a visit to Dilsberg,

Twain noted that he saw no idiots there, the captain explained that this was "Because of late the government has taken to lugging them off to asylums and otherwheres." Twain comments: "The captain probably imagined all this, as modern science denies that the intermarrying of relatives deteriorates the stock."[2]

This remained the majority opinion of "modern science" for the next seventy-five years, as is evident in the reaction to Edward Westermarck's suggestion that "the psychical cause" of the incest taboo "has a biological foundation in injurious consequences following unions of the nearest blood relatives."[3] Although Westermarck could quote in support of his suggestion the opinions of both Charles Darwin and Alfred Wallace,[4] he was widely accused of ignoring the findings of modern science. Lord Raglan faulted him for having assumed that inbreeding is harmful "in the face of all the evidence"[5]; Bronislaw Malinowski argued against him that "biologists are in agreement that there is no detrimental effect produced upon the species by incestuous unions"[6]; and Robert Briffault claimed that "there is not in the records of breeding from domesticated animals a single fact . . . which indicates, much less evidences, that inbreeding, even the closest, is itself productive of evil effects."[7]

In 1934 Westermarck rebutted these criticisms in detail, but his arguments were universally ignored. Writing fifteen years later, Leslie White had no doubts whatsoever about the relationship between inbreeding and the incest taboo. There is none. The theory that incest was prohibited because inbreeding causes biological deterioration is "so plausible as to seem self-evident, but it is wrong for all that. . . . Inbreeding as such does not cause degeneration; the testimony of biologists is conclusive on this point."[8] According to White, inbreeding can only intensify the inheritance of traits, good or bad. If Dilsberg was an idiot factory, it is only because the founders were inclined toward idiocy. In societies where brother-sister marriage is permitted in the ruling family, "we may find excellence. Cleopatra was the offspring of brother-sister marriages continued through several generations and she was 'not only handsome, vigorous, intellectual, but also prolific . . . as perfect a specimen of the human race as could be found in any age or class of society.'"[9]

Claude Lévi-Strauss, writing only a year after White, reached the same conclusion. He acknowledged E. M. East's work on maize and his view that "because objectionable recessive traits are as common in the human race as they are in maize,"[10] folk beliefs about the injurious effects of inbreeding are largely justified. But he then argued that "East's work has indirectly established that these supposed dangers would never have appeared if mankind had been endogamous from the beginning." His conclusion was that "the temporary danger of exogamous unions, supposing such a danger to exist, obviously stems from an exogamous or pangenetic tradition, but it cannot be the cause of this tradition."[11] Lévi-Strauss's preferred authority was George Dahlberg, who concluded that "as far as heredity is concerned these inhibitions [i.e., the incest taboos] do not seem to be justified."[12]

When Aberle et al. met at Stanford in 1956, they too appear to have taken the position that inbreeding is not necessarily deleterious. But by the time they published the results of their deliberations in 1963, they had changed their minds because of "new information" "which appeared after the 1956 argument had been developed."[13] This new information was, first, the finding that "the ratio of deleterious and lethal recessive genes to selectively advantageous genes is very high indeed," and, second, the finding that "the percentage of individuals homozygous for lethal or deleterious recessives rises sharply" as degree of relatedness increases. This led to the published conclusion that "the biological advantages of the familial incest taboo cannot be ignored." "Close inbreeding of an animal like man has definite biological disadvantages, and these disadvantages are far more evident as respects the mating of primary relatives than as respects other matings."[14] In other words, Westermarck (and Twain's captain) were right all along.

The present state of our knowledge of the consequences of inbreeding is summarized in this volume by Alan H. Bittles in Chapter 2. For Bittles, the question is not whether inbreeding is injurious; it is how injurious. Bittles's concern is that the rarity of sibling and parent-child unions combined with their being tabooed may have led researchers and makers of social policy to overestimate the dangers of close inbreeding. One possibility is that "a rigorous examination, including determination of paternity, may only be initiated if a child shows symptoms of physical and/or intellectual handicap." Another is that studies of familial incest do not have adequate controls for "the potentially adverse effects of nongenetic variables, such as very young or advanced maternal and paternal ages, and unsuccessful attempted interruption of the pregnancy."

Bittles's solution to these problems is to use cousin and uncle-niece unions to estimate the dangers of sibling and parent-child unions. The advantage of this method is that in many parts of the world—most notably Japan, South India, and Pakistan—cousin and uncle-niece unions are both legal and common. Consequently, we now have numerous studies (many conducted by Bittles himself) of the biological consequences of such marriages. Estimates based on these studies put the excess death rate among the progeny of sibling and parent-child unions at 16 to 20 percent, and the excess morbidity rate for such progeny at 6 to 16 percent. This suggests a total death and major disabilities rate of somewhere between 22 and 36 percent.

Although the "new information" that changed Aberle et al.'s minds indicated that inbreeding was at least as dangerous as Bittles's estimates imply, they rejected the possibility that primitive man recognized this danger.[15] This was a consequential decision, because it denied them a simple explanation of the incest taboo. They could not argue that it is a consciously instituted prophylactic. But how, then, did the taboo come into being? Probably because Eckhard H. Hess was a member of their workshop, Aberle et al. seriously considered the possibility that it expresses a disposition found in other species. Hess had shown that so long as they are reared together, Canada geese from the same brood never mate.

Experimental work . . . indicates that this fastidious behavior is the result of sexual imprinting. It is necessary to emphasize that the reaction persists without external sanctions. The luckless breeder who takes a male and female from the same brood to raise geese is doomed to disappointment: the pair will not mate even if no other partners are available. If, however, two members of the same brood are separated before hatching occurs and are subsequently re-introduced to each other, having been raised in different families, they may become mates.[16]

Although Aberle et al. failed to make the connection, this is evidence for what has long been known as "the Westermarck hypothesis." Westermarck argued that close inbreeding is injurious, but he did not argue (as Aberle et al. mistakenly imply) that recognition of the fact led to the incest taboo. Instead, he argued that the deleterious consequences of inbreeding have selected for an innate tendency to develop an aversion to sexual relations with childhood associates. This tendency, not recognition of the dangers of inbreeding, was the source of the incest taboo. As he put it in 1934 in what turned out to be his last words on the subject,

I must confess that the attempts to prove the harmlessness of even the closest inbreeding have not shaken my opinion that there is convincing evidence to the contrary. And here I find, as before, a satisfactory explanation of the want of inclination for, and consequent aversion to, sexual intercourse between persons who from childhood have lived together in that close intimacy which characterises the mutual relations of the nearest kindred. We may assume that in this, as in other cases, natural selection has operated, and by eliminating destructive tendencies and preserving useful variations has moulded the sexual instinct so as to meet the requirements of species.[17]

Might it not be, then, that Westermarck was right about the effects of early association as well as the dangers of inbreeding? Indeed, might it not be that Alfred Wallace was right in thinking that Westermarck had "solved the [incest] problem"?[18] Aberle et al. do not give the possibility a moment's consideration. They mention what they call "the indifference theory . . . only for the sake of completeness": "The indifference theory has both logical and empirical difficulties. It is hard to see why what is naturally repugnant should be tabooed, and the evidence for sexual attraction among kinsmen is quite adequate for rejecting the theory. We mention it only for the sake of completeness."[19]

The "logical difficulties" refer to Sir James Frazer's claim that the existence of the incest taboo is alone adequate to prove that Westermarck was wrong. In Sir James's words,

It is not easy to see why any deep human instinct should need to be reinforced by law. There is no law commanding men to eat or drink or forbidding them to put their hands in the fire. Men eat and drink and keep their hands out of the fire instinctively for fear of natural not legal penalties. . . . The law only forbids men to do what their instincts incline them to do; what nature itself prohibits and punishes, it would be superfluous for the law to prohibit and punish. Accordingly, we may always safely assume that crimes forbidden by law are crimes that many men have a natural propensity to commit. If there was no such propensity there would be no such crimes, and if no such crimes were committed what need to forbid them? Instead of assuming, therefore, from the legal prohibition of incest that there is a natural aversion to incest, we ought rather to assume that there is a natural instinct in favour of it, and that if the law represses it, as it represses other natural instincts, it does so because civilized men have come to the conclusion that the satisfaction of these natural instincts is detrimental to the general interests of society.[20]

This argument has been repeated, mantralike, by Westermarck's many critics. It was quoted in full by Sigmund Freud as early as 1911 and by Maurice Godelier as late 1989.[21] It was noted by critics as diverse in their views as George Peter Murdock, Claude Lévi-Strauss, Marshall Sahlins, and Richard Lewontin. It should be entered as exhibit 1 by anyone arguing that twentieth-century social thought was biased toward what I call functionalist fundamentalism.

The empirical difficulty Aberle et al. have in mind is Freud's claim that "psychoanalytic investigations have shown beyond the possibility of doubt that *an incestuous love choice* is in fact the first and regular one."[22] This is another anti-Westermarck mantra. It is repeated by A. L. Kroeber, Brenda Seligman, Leslie White, Alexander Goldenweiser, Claude Lévi-Strauss, and Marvin Harris, to name only the best-known of the many authors who quoted Freud against Westermarck. It is an essential part of functionalist fundamentalism because the view that human nature is selfish and unruly (perhaps even sinful) is necessary to the view that laws exist because they are needed.

Although Westermarck cited evidence suggesting that early association inhibits sexual attraction among many mammals (including horses and dogs), Aberle et al. ignored this and the experience of generations of animal breeders. It appears that they cited Hess only because it would have been impolite to ignore the work of an eminent colleague. Their unqualified conclusion was that "there is no evidence to suggest that asexual imprinting occurs among mammals." Asexual imprinting "does not seem to occur in man, the apes, the monkeys, or even in more remote mammalian species." We have therefore to assume that "this adaptive device was simply not available—not a part of the genetic equipment of man's ancestors or relatives."[23]

Although Aberle et al.'s conclusion regarding "more remote mammalian species" was unjustified, they can be excused for concluding that sexual imprinting does not seem to occur among the apes and monkeys. They wrote before primatology was an established research field. The difference this has made is dramatically summarized in Chapter 3, by Anne Pusey. After briefly reviewing evidence suggesting that inbreeding is injurious for most mammals (and more so in the wild than in captivity), Pusey catalogs a wealth of evidence indicating that something like asexual imprinting *is* found among our nearest relatives—rhesus macaques, baboons, gorillas, bonobos, and chimpanzees. Field and laboratory studies of "nonhuman primates provide abundant evidence for an inhibition of sexual behavior among closely related adults," *and* "the primate data support Westermarck's theory that familiarity during immaturity is a major reason for this avoidance." In several species immature mates do engage related females sexually, "but [this behavior] stops before the risk of conception."

The incest taboo posed a nearly impossible task for the functionalist fundamentalists. Rejecting Westermarck in favor of Freud, they had to find a supranatural source for the taboo. Their solution was to resurrect and remodel Edward Burnett Tylor's 1889 suggestion that "among tribes of low culture there is but one means of keeping up permanent alliances, and that is by means of intermarriage."[24] The essential first step in the argument was to insist that the incest taboo is only a way of implementing exogamy. "Nuer say that marriage to persons standing in certain relationships is forbidden because it is *rual*, incestuous," but E. E. Evans-Pritchard argued that "we may reverse this statement and say that sexual relations with persons standing in these relationships are considered incestuous because it would be a breach of the marriage prohibitions to marry them. I would hold that the incest taboo can only be understood by reference to the marriage prohibitions, and that these prohibitions must be viewed in the light of their function in the Nuer kinship system and in their whole social structure."[25]

Putting exogamy before the incest taboo led to the remarkable conclusion that the incest taboo is the means by which human beings transcended their animal nature. For Leslie White and Claude Lévi-Strauss, this made the incest taboo the passage between nature and culture. I put their formulations side by side to show how two authors who shared little else reached the same conclusion about the origins of the taboo. The similarity is evidence that they were responding to intellectual trends larger than themselves. First, Leslie White:

In the primate order . . . the social relationships between mates, parents and children, and among siblings antedates articulate speech and cooperation. They are strong as well as primary. And, just as the earliest cooperative group was built upon these social ties, so would a subsequent extension of mutual aid have to reckon with them. At this point we run squarely against the tendency to mate with an intimate associate. Cooperation *between* families cannot be established if parent marries child; and brother, sister. A way must be found to overcome this centripetal tendency with a centrifugal force. This way was found in the definition and prohibition of incest. If persons were forbidden to marry their parents or siblings they would be compelled to marry into some other family—or remain celibate, which is contrary to the nature of primates. The leap was taken; a way was found to unite families with one another, and social evolution as a *human* affair was launched upon its career. It would be difficult to exaggerate the significance of this step. Unless some way had been found to establish strong and enduring ties between families, social evolution could have gone no further on the human level than among the anthropoids.[26]

And then Claude Lévi-Strauss:

It will never be sufficiently emphasized that, if social organization had a beginning, this could only have consisted in the incest prohibition since, as we have just shown, the incest prohibition is, in fact, a kind of remodeling of the biological conditions of mating and procreation (which know no rule, as can be seen from observing animal life) compelling them to become perpetuated only in an artificial framework of taboos and obligations. It is there, and only there, that we find a passage from nature to culture, from animal to human life, and that we are in position to understand the very essence of their articulation.

As Tylor has shown almost a century ago, the ultimate explanation is probably that mankind has understood very early that, in order to free itself from a wild struggle for existence, it was confronted with the very simple choice of "either marrying-out or being killed-out." The alternative was between biological families living in juxtaposition and endeavoring to remain closed, self-perpetuating units, over-ridden by their fears, hatreds, and ignorances, and the systematic establishment, through the incest prohibition, of links of intermarriage between them, thus succeeding to build, out of the artificial bonds of affinity, a true human society, despite, and even in contradiction with, the isolating influence of consanguinity.[27]

Although *Les Structures élémentaires de la parenté* appeared in 1950 (and was summarized in English in an article published in 1956), Aberle et al. do not refer to Lévi-Strauss, but they do devote substantial space to White's version of what they call "the social and cultural systems theory" (which I prefer to call "group alliance theory"). In their view, "it is clear that the advantages postulated by White exist, and that, given a tendency to choose the most easily available mate, a complete prohibition on familial sexual relations is the simplest device for forcing ties between families." But they were not functionalists of the fundamentalist variety. They worried "that [White's] theory seems to assert that because the shift was advantageous, it came into being," and they worried that to come into being, the shift would "require a movement in opposition to certain strong trends."[28]

It requires the elimination of some younger members from the family, in spite of emotional attachments, and entrusting these members to groups where stable relationships do not yet exist. It also requires that primitive man understand the advantages of exchange—or else must assume that familial exogamy and the familial taboo arose as a chance "mutation" and survived because of their adaptive character.[29]

But what, then, could have motivated the change?

Aberle et al.'s premises put them in an awkward position. They accepted the evidence indicating that inbreeding is injurious, but they also accepted Freud's claim that human beings are naturally inclined to mate and marry within the family. Moreover, they rejected the idea that "primitive men" (whoever they might be) would have recognized either the disadvantages of inbreeding or the advantages of alliances. Having gotten themselves into this position by accepting Freud, they naturally turned to Freud to solve their problem. After his famous reconstruction of what he called the "emotional motive" for the incest taboo, Freud went on to argue that "it has a practical basis as well."

Sexual desires do not unite men but divide them. Though the brothers had banded together in order to overcome their father, they were all one another's rivals in regard to the women. Each of them would have wished, like his father, to have all the women to himself. The new organization would have collapsed in a struggle of all against all, for none of them was of such overmastering strength as to be able to take on his father's part with success. Thus the brothers had no alternative, if they were to live together, but—not, perhaps until they had passed through many dangerous crises—to institute the law against incest, by which they renounced the women whom they desired and who had been their chief motive for dispatching their father. In this way they rescued the organization which had made them strong.[30]

Citing Freud as the originator of what they call "the family theory" (and I call "group harmony theory"), Aberle et al. argued that the incest taboo was instituted to maintain order in the family. This was possible for an animal with "language and limited culture" because domestic strife would "be observable as a pressing problem, on a day-to-day basis, and the source of the problem in sexual competition would be equally evident." They recognized that the problem might be solved by regulating sex rather than by eliminating it, but argued that this solution would not survive over time because it would not solve the genetic problem posed by the dangers of inbreeding.[31]

Although Aberle et al. avoided many of the mistakes made by White and Lévi-Strauss, they ended up with a story that is no more convincing. It is, as Hill Gates asserts in Chapter 8 of this volume, the result of anthropology's "embrace . . . of Sigmund Freud's Oedipus complex and its odd alliance with contract theory." The incest taboo is taken as promising, if not utopia, at least a healthier, more orderly existence. But human nature, incestuously inclined, stands in the way. A means must be found to overcome it. Human beings must see that to fully realize their potential they have to repress their selfish sexual interests. It is a particularly attractive parable because everyone knows that the story has a happy ending. Somehow, somewhere, for some reason or other, our ancestors saw the light and made the necessary sacrifices. Rationality triumphed. All good men agreed that they would forgo their sisters and exchange them for wives.

The antihero of this heroic tale is Edward Westermarck. If early association were to be found to inhibit sexual attraction among humans beings as well as among geese, the plot would lose its dramatic motive and much of its appeal. Our ancestors would not have had to repress their natural inclinations to harvest the advantages of outbreeding and exogamy. They would be guaranteed by their natural inclinations. It is, then, ironic that even before Aberle et al.'s version of the story appeared in print, the Frazer/Freud tide had turned. In 1962 Robin Fox published an essay in which he argued that reaction to the possibility of sex among persons who have experienced close bodily contact as children "varies from 'disgusting' or 'unthinkable' to 'indifferent.'" "It is the reaction of indifference that we find most interesting, and most neglected, due to the facile rejection of Westermarck's observation."[32]

Fox's bellwether essay was followed two years later by Yonina Talmon's study of sexual relations among children reared together in two Israeli kibbutzim, four years later by my first report of the sexual consequences of minor marriages in Taiwan, and less than a decade later by Joseph Shepher's survey of a large sample of marriages in Israel. All three studies documented "a lack of inclination for . . . sexual relations between persons who have lived together in a long-continued relationship from a period of life when the actions of sexual desire, in its acuter forms at least, is naturally out of the question."[33]

A Westermarck revival was under way. In the years since, it has amassed evidence that leaves little doubt that Aberle et al. erred in concluding that asexual imprinting does not have an analogue among humans. The only ethnographic case that could ever be mustered in support of their conclusion—brother-sister marriage in Roman Egypt—is nullified by Walter Scheidel in Chapter 5 of this volume. After carefully reexamining the forty-six known cases of sibling and half-sibling marriages, he concludes that "all in all, there is nothing to show that as far as the correlation of early childhood association and sexual inhibition is concerned, the evidence for Roman

Egyptian sibling marriage deviates significantly from the pattern derived from the Chinese data on 'minor marriages' and other information on the demographic context of incestuous behavior and incest avoidance in humans."

Thus, the second half of the twentieth century saw two major changes in the assumptions researchers bring to the incest problem. The first is that close inbreeding is injurious. Denied by White in 1949, by Lévi-Strauss in 1950, and probably by Aberle et al. in 1956, the position advocated by Westermarck since 1890 was well on its way to general acceptance by 1963. The second and equally important change was the discovery that among most mammals and all the primates—including, most definitely, humans— early association inhibits sexual attraction. Again Westermarck was proved right. Thus the man who was mentioned in 1963 "only for the sake of completeness" enters the twenty-first century as almost the only man worth mentioning.

Recognition of the importance of these changes is what unites the chapters in this volume, but they are not the focal subjects of the volume. There are three problems raised by the discovery that inbreeding is injurious and early association inhibiting. I call them the mediation problem (A), the representation problem (B), and the localization problem (C). The mediation problem is, How are the deleterious consequences of inbreeding and the inhibiting nature of early association related? Most of the contributors to the volume are willing to assume that they are linked as cause and effect, but William Durham, volume editor and author of Chapter 7, demurs. He worries that "the Westermarck effect" may not be an adaptation. "What is lacking," in his view, "is conclusive evidence to show that the aversion was specifically shaped over time by genetic selection for the function it now performs."

I think doubts of this kind arise because we still do not know how the Westermarck effect is effected. In other words, we still do not know what causes us to respond to early association with an enduring aversion. I call the search for this cause the mediation problem because I am confident that whatever it is, it is a product of the dangers of inbreeding. It is what this selective force selected for. My hope is that when the mediation problem is solved the solution will convince skeptics like Durham that "in this case, as in other cases, natural selection has operated . . . so as to meet the requirements of the species."

When we solve the mediation problem (A), we will know why people avoid incest, but we will still not have an answer to the question that Bateson puts at "the heart of the matter": "What relations, if any, can be found between the avoidance of inbreeding and the incest taboo?" In other words, we will still have to solve the representation problem (B). Generally speaking, this is the problem of how the loves and hopes and fears and phobias of individuals give rise to norms, if they do. Bernard Williams (who was the first to use the term *representation problem*) puts it this way:

It is the notion of a *norm* that perhaps gives rise to the central representation problem. . . . The most, it seems, that a genetically acquired character could yield would be an inhibition against behaviours of a certain kind; what relation could that have to a socially sanctioned prohibition? Indeed, if the inhibition exists, what *need* could there be for such a prohibition? If a prohibitionary norm is to be part of the "extended phenotype" of the species, how could we conceive, starting from an inhibition, that this should come about?[34]

As I analyze it, the representation problem consists of a cluster of three related problems concerning the relationship between individual inclinations and social regulations. I call them the externalization problem (B1), the expression problem (B2), and the moralization problem (B3). The externalization problem concerns the fact that the incest taboo is not a matter of self-regulation. It is a matter of public condemnation. The fact that early association inhibits sexual attraction explains why most people avoid sexual relations with their parents and siblings, but it does not explain why they condemn other people for having sexual relations with *their* parent or *their* sibling.

Although it is implied by his general statement, Williams did not address the externalization problem (B1). For him, the core of the representation problem is what I call the expression problem (B2). And, in his view, it is an insoluble problem. For Williams there is, as Neven Sesardic puts it in Chapter 6, a "transcendental obstacle" blocking all movement from biologically based inhibitions to socially sanctioned prohibitions. I call it "the expression problem" because it concerns the conceptual content of the incest taboo. The argument is that the aversion aroused by early association cannot possibly explain the taboo because they are about different things. One is about "those [people] that one is brought up with"; the other is about "marriages that could constitute close inbreeding." In Williams's own words,

Not only does extra conceptual content have to be introduced to characterize the human prohibition, but also the introduction of that content stands in conflict with the proposed explanation of it. There are no sanctions against marrying those that one is brought up with (as such); the sanction is against marriages which would constitute close inbreeding. The conceptual content of the prohibition is thus different from the content that occurs in the description of the inhibition. It indeed relates to the suggested *function* of that inhibition, but that fact will not explain how the prohibition which is explicitly against inbreeding will have arisen. It certainly does not represent a mere "raising to consciousness" of the inhibition.

Although it is also implied by his general statement of the representation problem (B), Williams fails to address what I call the moralization problem

(B3). This is surprising given his interest in morality, because the problem arises from the fact that universally the incest taboo is represented as having strongly felt moral content. Not only do people disapprove of incest, their disapproval is accepted as morally motivated. A solution to what I call the externalization problem would explain why people condemn incest, but it would not explain why such condemnation elicits universal approbation.

Neither the mediation problem (A) nor the representation problem (B) were seriously discussed until the end of the twentieth century. Until the early 1970s this was because most researchers accepted White and Lévi-Strauss's contention that the source of the incest taboo was supranatural; afterward, it was because the succeeding generation of cultural relativists decided there was no incest taboo. Their reasons were succinctly stated by Rodney Needham as early as 1971. "I conclude," Needham wrote, "that 'incest' is a mistaken sociological concept and not a universal." There were, in Needham's view, two reasons for this conclusion:

The first is the wide and variable range of statuses to which the prohibitions apply. The scope of application is in each case an integral feature of the social system, and in some sense a function of it; i.e., the complex of prohibitions in a society cannot be comprehended except by a systematic purview of the institutions with which they are implicated. By this account of the matter there are as many different kinds of incest prohibitions as there are discriminable social systems.

The second consideration is that the incest prohibitions are in part moral injunctions; they are expressions of indigenous ethical doctrines and, whether or not they are touched with a peculiar emotional quality, they have cultural meanings which no attempt at explanation can reasonably neglect.[35]

Needham's arguments were quickly seconded by David M. Schneider, Peter Riviere, and Roy Wagner. In fact, arguments that were original with Needham in 1971 were orthodoxy by the end of the decade. Thus, my preceding account of the fate of Aberle et al. is clearly "a presentist" version of history. What I have called a Westermarck revival was only a minor countercurrent in a tide drawn by the view that there is no incest taboo at all, only clusters of cultural particulars. The history of the period would be better represented by focusing on David M. Schneider than on Edward Westermarck. Although Schneider was a member of the Aberle et al. group (and, I was once told, drafted the report they published), he soon followed Needham's lead and abandoned the incest taboo as "a mistaken sociological concept." In a paper published in 1976 his "main point" with respect to the problem of incest was "to stop looking for causal explanations of origin, functional explanations of maintenance and to start looking at it as a problem in meaning in its cultural context."[36]

Now that the high tide of cultural relativism is receding, we can see that though it tried, it did not succeed in dissolving all social phenomena into

sion is aroused only when "the idea of sexual relations with a near relative occupies the mind with sufficient intensity and a desire fails to appear."[38] The reason we condemn other people for having sex with their relatives is because it does just this to us. We condemn them because by arousing our aversion their behavior causes *us* pain.

William Durham and I teach together and often debate in class the merits of Westermarck's solution to the externalization problem. I recommend it; he rejects it. Ironically, as he sees it, the three studies that have done the most to confirm Westermarck's account of why we avoid incest all disconfirm his account of why we condemn incest. In Taiwan, Israel, and Lebanon, children who are not siblings were commonly reared together as intimately as if they were siblings. In all three cases the result was, as Westermarck predicted, an aversion to sexual relations as adults. But in no one of these cases did the aversion aroused by early association produce a tendency to condemn sexual relations between the co-reared children. The children reared together in Israeli kibbutzim were encouraged to marry, and the children reared together in Taiwan and Lebanon were condemned if they refused to marry.

Durham offers in place of Westermarck's solution to the externalization problem a solution of his own. This is a more sophisticated version of the view that Aberle et al. rejected when, accepting the dangers of inbreeding, they denied that primitive man could have recognized these dangers. Durham argues that in so doing they overlooked the evidence preserved in the origin myths of many societies. This evidence says that the deleterious consequences of inbreeding were widely recognized in prehistoric times. The incest taboo is not, as Westermarck would have it, an unintended consequence of our emotional constitution; it is a consciously implemented solution to a recognized problem. Durham does not deny Westermarck's claims with regard to the consequences of early association. He even agrees that this is the primary reason humans avoid incest. What he denies is that the "social fact" we call the incest taboo is largely a product of the aversion aroused by early association. Thus, for Durham, what Williams calls the representation problem is not a problem at all. The incest taboo is not a representation. It is a creation.

Durham's position is best seen in contrast to the positions taken by Bateson (Chapter 1) and Gates (Chapter 8). Although Durham is reluctant to attribute the inhibiting effects of early association to the dangers of inbreeding, he is happy to attribute the incest taboo to these dangers. Bateson and Gates, in contrast, seem willing to assume that the dangers of inbreeding account for the inhibiting effects of early association, but they are not willing to assume that these dangers account for the incest taboo. "In summary," Bateson writes, "I suggest that it is unlikely that inbreeding avoidance and incest taboos evolved by similar mechanisms or even have a common utility in modern life."

In Chapter 6, Neven Sesardic challenges Durham's rejection of Wester-marck's solution to the externalization problem by suggesting that Durham is taking unfair advantage of an "epistemological tension" in Westermarck's account of the incest taboo. Westermarck argues that early association arouses a sexual aversion that gives rise to the incest taboo. His critics argue that the aversion he identifies is a result of the incest taboo, not its source. Thus, to make his case, Westermarck must show that even in the absence of an incest taboo, early association arouses an aversion. Consequently, every case Westermarck can muster in support of his aversion hypothesis will inevitably challenge his solution to the externalization problem. "It is," as Sesardic puts it, "a zero-sum game; what the theory gains by collecting evidence in favor of the [first hypothesis] it automatically loses on the other front because the same empirical data chip away at the [second hypothesis]." Durham responds (in Chapter 7) by arguing that what is significant about these cases is not simply that they involve socially acceptable unions between childhood associates. It is that in all three cases they are the marriages most parents prefer.

Chapter 6 also takes up Bernard Williams's treatment of what I have called the expression problem (B2). Williams argued that Westermarck was attempting the impossible in trying to derive the incest taboo from the aversion aroused by early association. This is impossible because the content of the aversion and the content of the taboo are not the same. The aversion is about sexual relations with the people with whom one is reared, while the taboo is about marriages between people who are classified as close relatives.

Sesardic argues, in defense of Westermarck, that Williams's argument only "looks persuasive because it trades on a crucial ambiguity." The fact that childhood association arouses an aversion to sexual relations does not necessarily mean that the people affected experience the aversion as being directed to their childhood associates *qua* associates. We must distinguish the cause of the aversion and the subjective experience of the aversion. When we do so, Williams's objection evaporates. It is not only possible, but also highly likely because it would be culturally encouraged, that siblings who are reared together experience the aversion aroused by their early association in terms of kinship rather than in terms of association. Thus, it is also highly likely that there is rarely a mismatch between the *experience* of childhood association and the content of the incest taboo.

There is, in my view, a relationship between Sesardic's critique of Durham and his critique of Williams. He could have employed against Durham the same argument he employs against Williams. Both confuse the cause of sexual aversion and the experience of that aversion. In all three of the societies Durham cites to refute Westermarck, the great majority of the people with whom one associates as a child are parents and siblings. Thus it is likely that the aversions aroused by childhood association are typically experienced in kinship terms. There is, then, no reason to expect that marriages involving childhood associates who are not siblings will elicit disapproval. They lack what it takes to turn a comfortable indifference into a painful aversion.

Most authors recognize that the incest taboo has what George Peter Murdock called "a peculiar emotional intensity,"[39] but many do not recognize that it also has a peculiar moral intensity. The result is that what I call the moralization problem has been neglected. In fact, the only thorough treatment of the problem is in Westermarck's *Origin and Development of Moral Ideas*. His solution to the problem is deceptively simple. It is premised on the view that "the moral concepts, which form the predicates of moral judgements, are ultimately based on moral emotions, that they are essentially generalizations of tendencies in certain phenomena to call forth indignation or approval." What distinguishes the moral emotions from other emotions is "their disinterestedness, apparent impartiality, and flavour of generality."[40] Thus, what makes disapproval of incest moral is the fact that the disapproval is general and does not appear to serve any selfish interest. In sum, it is moral because it is generally approved disapproval.

In Chapter 10, Larry Arnhart points out that even those social and biological scientists who defend Westermarck's explanation of incest avoidance reject his solution to the moralization problem. With the notable exception of E. O. Wilson, they cannot accept the possibility that moral concepts "are ultimately based on moral emotions." Even the evolutionary psychologists who take Westermarck's aversion hypothesis as paradigmatic reject his evolutionary approach to ethics as violating a fundamental fact/value dichotomy. For David Buss and Steven Pinker, as for Callicles and Kant, *is* is *is* and *ought* is another thing.

Arnhart contrasts "the 'transcendentalist' claim that ethics is rooted in absolute standards that exist outside of the human mind" with "the 'empiricist' claim that ethics is rooted in natural human inclinations." The contrast is neatly illustrated by the difference between Francis Hutcheson and Bernard Mandeville on the subject of the incest taboo. Hutcheson, in the empiricist tradition, argued that the incest taboo shows that we are all possessed of an innate moral sense. "Had we no *moral Sense natural* to us, we should only look upon *Incest* as hurtful to ourselves, and shun it, and never hate other incestuous Persons, more than we do a *broken Merchant*."[41] Mandeville, in the transcendentalist tradition, emphatically denied the existence of an innate moral sense. He agreed that "incestuous alliances are abominable; but it is certain that, whatever Horror we conceive at the thought of them, there is nothing in Nature repugnant against them, but what is built upon Mode and Custom."[42]

Arnhart's contrasts help us to understand why Westermarck's solution to the moralization problem was neglected. The reason is that just as Westermarck was a thoroughgoing Darwinian, so also was he—ipso facto, I would say—a thoroughgoing empiricist. Recognizing the moral content of the incest taboo, he refused to separate it from the taboo's psychological roots and its biological origins. Instead, he attributed the moral content of the taboo to the same sources as the aversion responsible for incest avoidance. This violated the fact/value dichotomy by turning an *ought* into an *is*. The argument would have appealed to Francis Hutcheson. It did not appeal to the transcendentalists who dominated twentieth-century social thought.

It is necessary to partition the incest problem and name the parts lest we lose sight of its complexity and be tempted by partial solutions. We must always remember, however, that solutions to one part of the problem may have implications for our understanding of other parts. This is illustrated over and over again in this volume. The implications of Bittles's concern that the immediate effects of inbreeding have been exaggerated are not limited to the problem of how incest avoidance originated. They are also relevant to the problem of why incest avoidance was mandated. The less visible the effects of inbreeding among a couple's children, the more likely it is that Westermarck was right in seeking the origins of the incest taboo among the passions. The more visible the effects of inbreeding among a couple's children, the more likely it is that Durham is right in arguing that rationality played a critical role.

One of the questions I take up in my own chapter is whether men are as sensitive to the inhibiting effects of early association as women are. Havelock Ellis suggested that they are not. I offer evidence indicating that he was wrong. Which of us is right will affect not only the solution to the mediation problem but, even more critically, the solution to the representation problem. It is one thing to derive a prohibition that applies to both sexes from an aversion felt by both, quite another to derive a prohibition that applies to both sexes from one felt by only one sex, particularly if the inhibited sex is also the socially subordinate sex.

Hill Gates concludes Chapter 8 by noting that "under conditions not fully mapped out, but surely recurrently, our innate alertness to the emotional complexity of incest was seized upon and turned to precise cultural ends, until something better came along." She is pointing to the possibility of relationships between the mediation problem and the localization problem. Mark Erickson and I argue that it is not coincidental that we sexually avoid those persons to whom we are most strongly attached as children.

Avoidance and attachment have evolved together, and together from the emotional core of the parent-child relationship. Thus, from our point of view, it is easy to see why secular and religious leaders who set themselves up as "father and mother of the people" extend the scope of the incest taboo and enact severe sanctions against incest. Sanctioning incest fits emotionally—and thereby helps justify politically—a paternalistic stance.

Their published report suggests that, meeting in 1956, Aberle et al. did not concern themselves with the practical or policy aspects of the incest problem. In my view, this was not because there was not yet an "incest epidemic," or that, if there was, it had not yet been diagnosed. It was because in 1956 incest was considered, at worst, a legal or ethical problem. In the eyes of scientists (social and biological alike), it did not have a medical or psychopathological aspect. Incest was a social matter without serious implications for either physical or mental health. It could be safely left in the hands of anthropologists with no interest in either biology or psychology.

The chapters in this volume show how Westermarck's return has changed all this. Bittles's interest in calculating as precisely as possible the level of defects among children of consanguineous unions is not entirely academic. It is motivated by practical concerns and has clear policy implications. Although the frequency of cousin and uncle-niece marriages has declined worldwide, it is still high in parts of Asia and Africa and in many immigrant communities in Europe. And there would be reason for concern even if the frequency of consanguineous marriages was lower than it actually is. An important part of Bittles's argument is that as infant mortality and deaths owing to infectious diseases decline, genetic disorders will constitute an ever larger proportion of the medical problems people experience. An inevitable result will be that the problems produced by consanguineous unions will be ever more obvious and ever more likely to stigmatize those people whose customs encourage such unions. He is particularly concerned about the effects of these changes in Europe, where these people are immigrant minorities.

Although their importance was long obscured by the mistaken views of Twain's "modern science," questions about the medical consequences of incest have a long history. They were, as Bittles observes, the subject of intense debate in the 1850s and 1860s, leading to Sir John Lubbock's attempt to have questions about cousin marriages included in the 1871 census. What is new from a practical perspective is concern about the psychopathological consequences of incest. Mark Erickson begins Chapter 9 by noting that "a distinct recollection of my psychiatric training in the 1980s is that of a group of patients, mostly female, who presented such a bewildering array of symptoms as to defy diagnosis." In 1980 it had not yet been discovered that they were victims of incest.

Although we are a long way from understanding why incest predisposes people to psychopathologies, it now appears certain that it does. Erickson mentions, among the many problems of incest victims, major depression, alcohol and drug dependence, self-mutilation and suicide, and a wide range of stress-related illnesses. The evidence of marriages worldwide rule out the possibility that the cause of such a panoply of disorders is the experience of sex with an older man. What is critical is that the older man is a father or older brother. But why does this matter? What makes sex with a father or an older brother so disturbing? The answer may be implicit in the solution Erickson and I offer to the mediation problem. It is the symbiotic relationship of incest avoidance and attachment behavior. The evolutionary purpose of attachment is to elicit caretaking. This is its innate promise. Consequently, sexual advances by a father or brother are disturbing because they deny the promise of caretaking. They threaten abandonment. Victims of paternal incest commonly complain that their father "betrayed them" or "violated their trust." What they are saying, we suggest, is that a person who had promised care behaved in a way that denied having ever made such a promise.

Whether or not the specifics of this hypothesis prove correct, it is a step in the right direction because it assumes that incest is unnatural. Incest is, as Erickson puts it, "a kinship pathology." So long as social scientists accepted the mid-twentieth-century view that human beings were naturally incestuous, the problems Erickson lists could not be reasonably attributed to incest. Incest was, in the twentieth-century view, a purely social delict. It could arouse, at worst, feelings of guilt or shame, and then only if the delict were discovered. It did not have the emotional stuff needed to fuel a major pathology.

Erickson concludes Chapter 9 by putting his subject in historical perspective: "During most of the twentieth century, social scientists believed incest was common in nature. Among humans, incest was thought to be rare because of cultural taboos." Ironically, Erickson suggests, "this view has been turned on its head. Incest is now known to be rare in nature, and we must seriously ask if human incest has become more, not less common, *because* of cultural influences." Thus, in Erickson's view, the twentieth century has seen a complete revolution in the relationship between culture, incest, and human welfare. Culture has gone from what saves us from our incestuous inclinations to what exposes us to the dangers of incest. What early in the century was seen as a need to repress our innate inclinations is now seen as a need to recover them.

What will we see in the twenty-first century? In Chapter 10 Larry Arnhart suggests that "we will see great advances in the biological understanding of human nature. He continues:

This will force us to think about whether biological science can explain that most distinctive trait of our humanity—our moral sense of right and wrong. Some people will argue that our moral experience transcends our biological nature. Others will argue that we should be able to explain our morality as an expression of our biological nature. How we decide that debate might be decisively influenced by whether we accept or reject Edward Westermarck's Darwinian theory of the incest taboo as a natural expression of human moral emotions."

I agree with Arnhart's assumption that the incest problem will be as hotly debated in the twenty-first century as it was in the twentieth century, and I agree with his prediction that the focal point of the debate will shift from the question of why we avoid incest to the question of why we condemn incest. I would like to add only the prediction that with this shift, the debate will become even more intense. I say this because in my view the underlying question that sustains interest in the incest problem is the degree to which we have managed to transcend our animal origins. Almost completely, argued the mid-twentieth-century transcendentalists, claiming the incest taboo as evidence that we are capable of overcoming, even remaking our animal nature. Then came the Westermarck revival showing that we avoid incest for natural, not cultural reasons. Now all that remains of the transcendentalists' case is the fact that we do not just "look upon *Incest* as hurtful to ourselves, and shun it." We "hate other incestuous Persons." This is the is/ought barrier, the transcendentalists' last line of defense. They will contest any attempt to breach it and do so as ardent champions of human dignity.

NOTES

1. David F. Aberle, Urie Bronfenbrenner, Eckhard H. Hess, Daniel R. Miller, David M. Schneider, and James N. Spuhler, "The incest taboo and the mating patterns of animals," *American Anthropologist*, vol. 65 (1963), p. 253.

2. Mark Twain, *A Tramp Abroad* (1880), (Oxford: Oxford University Press, 1996), p. 173.

3. Edward Westermarck, "Recent theories of exogamy," in *Three Essays on Sex and Marriage* (London: Macmillan, 1934), p. 147.

4. See Arthur P. Wolf, *Sexual Attraction and Childhood Association: A Chinese Brief for Edward Westermarck* (Stanford, Calif.: Stanford University Press, 1995), pp. 2–8.

5. Lord Raglan, *Jocasta's Crime* (London, 1933), p. 16.

6. Bronislaw Malinowski, *Sex and Repression in Savage Society* (London, 1927), p. 243.

7. Robert Briffault, *The Mothers* (London, 1927), vol. 1, p. 215.

8. Leslie White, "The definition and prohibition of incest," *American Anthropologist*, vol. 50, part I (1948), p. 417.

9. Ibid.

10. E. M. East, *Inbreeding and Human Affairs* (New York, 1938), p. 156.

11. Claude Lévi-Strauss, *The Elementary Structures of Kinship* (Boston: Beacon Press, 1969), trans. James Harle Bell, John Richard von Strummer, and Rodney Needham, pp. 14–15.

12. G. Dahlberg, "On rare defects in human populations with particular regard to inbreeding and isolate effects," *Proceedings of the Royal Society of Edinburgh*, vol. 58 (1937–38), p. 224.

13. Aberle et al., "Incest taboo," p. 256.

14. Ibid., pp. 256–57.

15. Ibid., p. 257.

16. Ibid., p. 259.

17. Westermarck, "Recent theories of exogamy," pp. 158–59.

18. Alfred Wallace, K. Rob, and V. Wikman, "Letters from Edward B. Tylor and Alfred Russell Wallace to Edward Westermarck," *Acta Academia Aboensis Humanioria*, vol. 13, no. 8 (1940), p. 18.

19. Ibid., p. 258.

20. J. G. Frazer, *Totemism and Exogamy* (London, 1910), vol. 4, pp. 97–98.

21. See Maurice Godelier, "Incest taboo and the evolution of society," in *Evolution and Its Influence*, ed. Alan Grafen (Oxford: Clarendon Press, 1989), p. 69.

22. Sigmund Freud, *A General Introduction to Psychoanalysis* (1920), trans. Joan Riviere (New York: Pocket Books, 1953), pp. 220–21.

23. Ibid., p. 261.

24. Edward B. Tylor, "On a method of investigating the development of institutions; applied to laws of marriage and descent," *Journal of the Royal Anthropological Institute of Great Britain and Ireland*, vol. 18 (1889), p. 267.

25. E. E. Evans-Pritchard, *Kinship and Marriage Among the Nuer* (London: Oxford University Press, 1951), pp. 43–44.

26. White, "Definition and prohibition," p. 425.

27. Claude Lévi-Strauss, "The family," in *Man, Culture, and Society*, ed. Harry L. Shapiro (New York: Oxford University Press, 1960), p. 278.

28. Aberle et al., "Incest taboo," p. 258.

29. Ibid.

30. Sigmund Freud, *Totem and Taboo*, trans. James Strachey (New York: W. W. Norton, 1950), p. 144.

31. Aberle et al., "Incest taboo," p. 262.

32. Robin Fox, "Sibling incest," *British Journal of Sociology*, vol. 13 (1962), p. 147.

33. Edward Westermarck, *A Short History of Human Marriage* (London, 1926), p. 86.

34. Bernard Williams, "Evolution and ethics," in *Evolution from Molecules to Men*, ed. D. S. Bendall (Cambridge: Cambridge University Press, 1983), p. 560.

35. Rodney Needham, "Remarks on the analysis of kinship and marriage," in *Rethinking Kinship and Marriage* (London: Tavistock, 1971), pp. 25–26.

36. David M. Schneider, "The meaning of incest," *Journal of the Polynesian Society*, vol. 85 (1976), p. 168.

37. Jude Cassidy and Phillip R. Shaver, eds., *Handbook of Attachment Behavior* (New York: Guildford Press, 1999).

38. Westermarck, *History of Human Marriage*, vol. 2, p. 197 or 214.

39. George Peter Murdock, *Social Structure* (New York: Macmillan, 1949), p. 288.

40. Edward Westermarck, *The Origin and Development of Moral Ideas* (London: Macmillan, 1906–8), vol. 2, pp. 738–39.

41. Francis Hutcheson, *An Inquiry into the Original of Our Ideas of Beauty and Virtue* (London: J. Darby for Will and John Smith, 1725), p. 192.

42. Bernard Mandeville, *The Fable of the Bees* (Oxford: Clarendon Press, 1924), vol. 1, p. 331.

1 *Inbreeding Avoidance and Incest Taboos*

Patrick Bateson

I have never much liked the way some of my colleagues in the biological sciences have applied terms such as *rape* or *marriage* to animals. I appreciate that this is sometimes done to lighten the normally dull language of scientific discourse. However, these terms have established usage in human institutions with all their associated rights and responsibilities of individuals and culturally transmitted rules on what people can and cannot do. Problems of communication between disciplines are compounded when, having found some descriptive similarities between animals and humans and having investigated the animal cases, biologists or their popularizers use the animal findings to "explain" human behavior. Such arguments rely on a succession of puns, which are usually unconscious, but which are especially unfunny to those social scientists who feel threatened by an apparent takeover bid of the biologists.

I believe that *incest* should be restricted to human social behavior where culturally transmitted proscriptions limit sexual contact and marriage with close kin (and others who might be deemed to be close kin). *Inbreeding avoidance* should be used for behavior that makes matings with close kin less probable in both humans and nonhuman animals. This separation then leaves open the question of whether these behaviors have evolved for similar reasons and whether the two phenomena have similar current functions.

This chapter briefly reviews the evidence that people unconsciously choose mates who are a bit different from those individuals who are familiar from early life but not too different. In a biological context this is often referred to as optimal outbreeding.[1] Why did it evolve? The question in-

Generous financial support from the Australian Research Council and the benefits of ongoing collaboration with many colleagues in Australia, India, Pakistan, the United Kingdom, and the United States is gratefully acknowledged.

vites examination of the concept of adaptation and the role of Darwinian evolution in generating such adaptations. Since evolution is thought to involve changes in genes, it is necessary to be clear about the role of genes in an individual's development. When development is considered, a quite different set of issues is raised. These need to be considered in relation to the formation of mating preferences. Finally, it is necessary to come to the heart of the matter: what relations, if any, can be found between the avoidance of inbreeding and incest taboos?

Optimal Outbreeding

The biological costs of inbreeding are evident enough in other animals.[2] They are particularly obvious in birds. If a male bird is mated with his sister, and their offspring are mated together, and so on for several generations, the line of descendants usually dies out fairly quickly. This happens because some damaging genes are more likely to be expressed in inbred animals. Some potentially harmful genes are recessive and therefore harmless when they are paired with a dissimilar gene, but they become damaging in their effects when combined with an identical gene. They are more likely to be paired with an identical recessive gene as a result of inbreeding. The presence of such genes is a consequence of the mobility of the birds and the low probability that they will mate with a bird of the opposite sex that is genetically similar to them. Over time, the recessive genes have accumulated in the genome because they are normally suppressed by their dominant partner gene.

The genetic costs of inbreeding arising from the expression of damaging recessive genes are the ones that people usually worry about. However, recessive genes are less important in mammals than they are in birds because mammals generally move around less and may live in quite highly inbred groups. The most important biological cost of excessive inbreeding is that it negates the benefits of the genetic variation generated by sexual reproduction. If an animal inbreeds too much, it might as well make many copies of itself without the effort and trouble of courtship and mating.

On the other side, excessive outbreeding also has costs.[3] For a start, excessive outbreeding disrupts the relation between parts of the body that need to be well adapted to each other. The point is illustrated by human teeth and jaws. The size and shape of teeth are strongly inherited characteristics. So too are jaw size and shape, as may be seen in the famous paintings of the Hapsburg family, scattered around the museums of the world. The Dürer painting of the Holy Roman Emperor Maximilian I reveals the large Hapsburg jaw, which remained as pronounced in his great-great-

great-grandson, Philip IV of Spain, shown in the painting by Velasquez. The potential problem arising from too much outbreeding is that the inheritance of teeth and jaw sizes are not correlated. A woman with small jaws and small teeth who had a child by a man with big jaws and big teeth lays down trouble for her grandchildren, some of whom may inherit small jaws and big teeth. In a world without dentists, ill-fitting teeth were probably a serious cause of mortality. This example of mismatching, which is one of many that may arise in the complex integration of the body, simply illustrates the more general cost of outbreeding too much.

Some of the evolutionary pressures on mate choice arose from too much inbreeding, on the one hand, and from too much outbreeding on the other. A preference for an individual somewhat like close kin will minimize the opposing ill effects of breeding with individuals who are genetically too different. A sexual preference for individuals who are a bit different from close kin strikes a balance between the biological costs of inbreeding and those of outbreeding.[4]

The suggestion is that individuals had greatest reproductive success if they mated with a partner who was somewhat similar to themselves, but not too similar. The hypothesis has gathered considerable empirical support from studies of animals.[5] Japanese quail, for example, prefer mates that are first or second cousins, when given a choice in laboratory experiments.[6] If they have been reared with unrelated individuals, the quail prefer mates that are a bit different from these familiar individuals. In humans a great mass of data shows that freely chosen human spouses are more like each other than would be expected on a chance basis. Similarities are not only social and psychological but also found in measures of body dimensions such as length of earlobe.[7]

Humans choose partners somewhat like themselves.[8] At the same time, people prefer sexual partners who look slightly different from individuals with whom they have grown up. Faces are perceived as more attractive if some of the facial features are exaggerated by caricaturing the image so that it differs from the average.[9] Most people are attracted to faces that are distinctive and depart from the average. When the faces of individuals who were perceived as being attractive were averaged, this composite was preferred over the average of all faces.

Natural experiments have been performed unwittingly on human beings. Famously, Israeli kibbutzniks grow up together like siblings and rarely marry each other.[10] The most comprehensive evidence has come from Arthur Wolf's (1994) study of the marriage statistics from Taiwan in the late nineteenth and early twentieth century, when Taiwan was under Japanese control.[11] The Japanese kept detailed records for the births, marriages, and deaths of everyone on the island. As in many other parts of China, marriages were arranged and occurred mainly and most interestingly in two forms. The major type of marriage was the conventional one in which the partners first met each other when adolescent. In the minor type of marriage, the wife-to-be was adopted as a young girl into the family of her future husband. In minor-type marriages, therefore, the partners grew up together like siblings. In this sense they were like the quail in the laboratory experiment, having been reared with an individual of the opposite sex to whom they were not genetically related. Later in life their sexual interest in their partner was assessed in terms of divorce, marital fidelity, and the number of children produced. By all these measures, the minor marriages were conspicuously less successful than the major marriages. Typically, the young couples who had grown up together from an early age, like brother and sister, were not much interested in each other sexually when the time came for their marriage to be consummated.

While a great deal is still unknown about sexual preferences in both animals and humans, the similarities are quite striking. The processes involved in preference of humans for slight novelty have been subject to Darwinian evolution. However, acceptance of this point has to be tempered by an awareness that mate choice is influenced by many qualities that are beyond the scope of this chapter.[12]

Adaptation and Darwinism

Unconscious preferences for slight novelty are seen by biologists as being adaptive in the sense that they serve to enhance the reproductive success of the individual who acts on those preferences. Darwinism has generated much distrust in the social sciences because it seemed to spawn such strange and, indeed, wicked social theories. The reason why biologists like me are still greatly enamored of the Grand Old Man is because he provided what is probably the only coherent and systematic explanation for adaptation—the match between the characteristics of organisms and the worlds in which they live.

Complicated things found in biology have the appearance of having been designed for a purpose. In the early nineteenth century a famous theologian, William Paley, put it this way: "It is the suitableness of these parts to one another; first, in the succession and order in which they act; and, secondly, with a view to the effect finally produced."[13] Paley took this as proof of the existence of God. Darwin provided us with an explanation of how it came about. When individual differences are inherited, those individuals that are better adapted than others are more likely to survive and reproduce and then have offspring that share their adaptations.

The perception that behavior is designed springs from the relations between the behavior, the circumstances in which it is expressed, and the resulting consequences. The closeness of the perceived match between the tool and the job for which it is required is relative. In human design, the best that one person can do will be exceeded by somebody with superior technology. If you were on a picnic with a bottle of wine but no corkscrew, one of your companions might use a strong stick to push the cork into the bottle. If you had never seen this done before, you might be impressed by the selection of a rigid tool small enough to get inside the neck of the bottle. The tool would be an adaptation of a kind. Tools that are better adapted to the job of removing corks from wine bottles are available, of course, and an astonishing array of devices have been invented. One ingenious solution involved a pump and a hollow needle with a hole near the pointed end; the needle was pushed through the cork and air was pumped into the bottle, forcing the cork out. Sometimes, however, the bottle exploded and this tool quickly became extinct. As with human tools, what is perceived as good biological design may be superseded by an even better design, or the same solution may be achieved in different ways.

Among those who spin stories about biological design, a favorite figure of fun is an American artist, Gerald Thayer.[14] He argued that the purpose of the plumage of all birds is to make detection by enemies difficult. Some of the undoubtedly beautiful illustrations were convincing examples of the principles of camouflage. However, among other celebrated examples, such as pink flamingos concealed in front of the pink evening sky, was a painting of a peacock with its resplendent tail stretched flat and matching the surrounding leaves and grass. The function of the tail was to make the bird difficult to see! Ludicrous attributions of function to biological structures and behavior have been likened to Rudyard Kipling's *Just-So* stories of how, for example, the leopard got his spots.[15] However, the teasing is not wholly justified. Stories about current function are not about how the leopard got his spots, but what the spots do for the leopard now. That question is testable by observation and experiment.

Not every speculation about the current use of a behavior pattern is equally acceptable. Both logic and factual knowledge can be used to decide between competing claims. Superficially attractive ideas are quickly discarded when the animal is studied in its natural environment. The peacock raises his enormous tail in the presence of females, and he molts the cumbersome feathers as soon as the spring breeding season ends. If Thayer had been correct about the tail feathers being used as camouflage, the peacock should never raise them conspicuously and he should keep them year round.

Genes and Development

A Darwinian account tells us nothing about how those sexual preferences develop in the individual. It is true that for the Darwinian evolutionary mechanism to work, something must be inherited. But even if a single gene provides the basis for the distinctive beneficial character of the individual, a single gene is not sufficient for the development of the character. This is where we get to the heart of a very lively debate in biology.

Scientists collaborating on the Human Genome Project have elucidated nearly all the DNA sequences of all the genes on all twenty-three pairs of chromosomes found in every human cell. It is a staggering achievement. But the excitement about what is being done should be greatly moderated. "The Book of Life," as one leading scientist called it, will not provide the complete story about human nature.

The human genome is like a cook's larder list.[16] Working out all the dishes that cooks might make from the ingredients available to them is another matter. If you want to understand what happens in the lifelong process from conception to death, you must study the process. The starting points of development include the genes. But they also include factors external to the genome, and of course, the social and physical conditions in which the individual grows up are crucial.

The language of a gene "for" a particular characteristic is exceedingly muddling to the nonscientist—and, if the truth be told, to many scientists as well. What the scientists mean (or should mean) is that a genetic difference between two groups is associated with a difference in a characteristic. They know perfectly well that other things are important and that, even in constant environmental conditions, the outcome depends on a combination of many genes. Particular combinations of genes have particular effects, and a gene that fits into one combination may not fit into another. Unfortunately, the language of a gene "for" a characteristic has a way of seducing the scientists themselves into believing their own sound bites. The language rests on a profound misunderstanding.

While genes obviously matter, even a cursory glance at humanity reveals the enormous importance of each person's experience, upbringing, and culture. Nobody could seriously doubt the remarkable human capacity for learning from personal experience and from others. It is obvious that experience, education, and culture make a big difference, whatever an individual's genetic inheritance. Individuals are not like the dry Japanese paper flowers that are simply put into water to open out.

The notion that genes are simply blueprints for an individual human is hopelessly misleading. In a blueprint, the mapping works both ways: starting from a finished house, the room can be found on the blueprint, just as the room's position is determined by the blueprint during the building process. This straightforward mapping is not true for genes and individual human behavior patterns, in either direction.

Genes do not make behavior patterns or physical attributes. Genes make proteins. Each human has about 30,000 genes, each of which is an inherited molecular strand (or set of strands) that may be translated into a protein molecule (or part of one). The proteins are crucial collectively to the functioning of each cell in the body. Some proteins are enzymes, controlling biochemical reactions, while others form the physical structures of the cell. These protein products of genes work not in isolation but in a cellular environment created by the conditions of the local environment and by the expression of other genes. Each gene product interacts with many other gene products.

Any characteristic of an individual, such as a behavior pattern or psychological attribute, is affected by many different genes, each of which contributes to the variation between individuals. In an analogous way, many different design features of a motor car contribute to a particular characteristic such as its maximum speed. A particular component such as the system for delivering fuel to the cylinders may affect many different aspects of the car's performance, such as its top speed, acceleration, and fuel consumption. A broken wire can cause a car to break down, but this does not mean that the wire by itself is responsible for making the car move.

The image of a genetic blueprint also fails because it is too static, too suggestive that adult organisms are merely expanded versions of the fertilized egg. In reality, developing organisms are dynamic systems that play an active role in their own development. To some extent each individual chooses and shapes its own physical and social environment. This can have interesting consequences. People who differ in ways that relate to differences in their genes may also pick certain physical and social environments in which to live. This process has been given the name "niche-picking."[17] It means that individuals with different characteristics, some of which reflect differences in their genes, end up by their own actions experiencing the world in different ways.

Environmental and inherited factors often work together to produce much larger overall effects than when either factor is present on its own. The often uncanny similarities between identical twins provide striking evidence for the importance of genes in shaping physical and behavioral characteristics. But one surprising finding to emerge from studies of identical twins is that twins reared apart are sometimes more like each other than those reared together.[18] To put it another way, rearing two genetically identical individuals in the same environment can make them less similar. This fact pleases neither the extreme environmental determinist nor the extreme genetic determinist.

The environmental determinist supposes that twins reared apart must have different experiences and should therefore be more dissimilar in their behavior than twins who grew up together in the same environment. The genetic determinist does not expect to find any behavioral differences between genetically identical twins reared together; if they have had the same genes and the same environment, then how can they be different? Of course, one twin provides a social environment for the other and siblings often hate to do what the other one is doing.

A single developmental ingredient, such as a gene or a particular form of experience, might produce an effect on behavior, but this knowledge gives only a feeble insight into developmental processes. The best that can be said of the nature/nurture split is that it provides a framework for uncovering a few of the genetic and environmental ingredients that generate differences between people. At worst, it satisfies a demand for simplicity in ways that are fundamentally misleading.

The processes involved in behavioral and psychological development have certain metaphorical similarities to cooking, to which I have already alluded. Both the raw ingredients and the manner in which they are combined are important. Timing also matters. In the cooking analogy, the raw ingredients represent the many genetic and environmental influences, while cooking represents the biological and psychological processes of development. Nobody expects to find all the separate ingredients represented as discrete, identifiable components in a soufflé. Similarly, nobody should expect to find a simple correspondence between a particular gene (or a particular experience) and particular aspects of an individual's behavior or personality.

The Development of Mating Preferences

Returning to mating preferences, how do they develop? A preference for faces that are a bit different, but not too different, from a familiar standard is relevant to mate choice in other species. Animals of many species tend to avoid mating with individuals who are very close kin, such as siblings, but they do sometimes prefer to mate with more distant relatives. The developmental process involved was first made famous by Konrad Lorenz and called "sexual imprinting."[19] This is coupled with some habituation to the very familiar, which offsets the preference to the slightly novel.[20]

The few Israeli kibbutzniks who chose to marry within their peer group were usually those who had entered the kibbutz after the age of six and therefore had not grown up with their future spouses.[21] In Taiwan, girls who were adopted into families before the age of three and then married their adopted "brother" had a lower fertility than girls adopted later.[22] Neither of these findings means that the learning process that affects adult sexual preferences is completed early in life. If children grow up together and, as a result, see a lot of each other, they revise the ways in which they recognize each other; this goes on until they are sexually mature. By the time they are three, children are highly conscious of their own sex and are much less likely to play with somebody of the opposite sex, particularly a child who is not well known to them.[23] It seems plausible then that a girl who is adopted when over three will be viewed as a stranger by the boy and treated differently from a girl who is adopted when younger.

Two developmental explanations have been proposed for the age-dependent effect. The first suggests that all the information that will be used is gathered within a period when the brain is most easily affected by such experience.[24] This is a classical critical period hypothesis used at one time for behavioral imprinting in birds but now largely discarded. It supposes that the adult can remember the faces of its siblings and generalize from their childlike characteristics to their adult form. As far as I know, this possibility has never been tested. I don't find it especially plausible, but in its defense, it should be remembered that algorithms have been devised for aging pictures of children so that those who have been kidnapped early in life might be recognized many years later. If computers can do it, then perhaps humans can too.

I prefer a rather different hypothesis, which is that while the process of learning starts most readily at a relatively early stage in development, the representations of familiar faces are updated by continuing close contact.[25] A lot hinges on the type of relationship the couple have when they are young. If they play together and, as a result, see a lot of each other, the indifference is likely to be greater. By the time they are three, children are highly conscious of their gender and are much less likely to play with a member of the opposite sex, particularly a strange member.[26] It seems plausible that a girl adopted when over three will be seen as a stranger by the boy and treated very differently from girls who are adopted when younger.

The idea that familiarity of a certain kind does reduce sexual attractiveness may be applied rewardingly to explain one striking feature of divorce statistics. For instance, in British women who married before the age of twenty, the proportion of marriages that ended in divorce has been approximately double that of the marriages of women who married between twenty and twenty-four.[27] The difference is found at any time between four and twenty-five years after marriage. Many factors, such as differences between social classes in attitudes toward marriage, could explain or contribute to explaining the difference. However, early marriages may involve a great deal of intimacy but relatively little sexual satisfaction. Indeed, people often report that their early sex lives were relatively unrewarding. If the effects of habituation are not powerfully offset by rewarding sexual experience, the partner may lose his or her attractiveness and become the equivalent of a sibling.

Evolution and Function of Mating Preferences

The presumption made by biologists is that inbreeding avoidance evolved because those individuals that did it were more likely to have greater reproductive success than those who did not. Since mate choice is affected by early experience in a great many birds and mammals, with a preference for types that are familiar from early life, it would not be a big step to superimpose on such a mechanism a reduced preference for those individuals that are very familiar. Two well-known learning processes, behavioral imprinting and long-term habituation, are able to generate a preference for individuals who are a bit different but not too different from those individuals who are familiar from early life.[28] If, as the evidence strongly suggests, inbreeding and excessive outbreeding carry biological costs in the form of reduced reproductive success, then the activation of both processes in the development of sexual preferences would have been favored. It does not matter for the evolutionary argument whether behavioral imprinting came before, after, or at the same time as the habituation of sexual preferences. Just how the early experience is mediated through the rest of childhood to affect adult sexual preferences remains unresolved. But uncertainty about the developmental argument does not detract from the evolutionary hypothesis that it was more beneficial to base sexual inhibitions on early experience rather than later experience. Those individuals who set the standards determining with whom they should not mate when they were likely to be surrounded by family were fittest. They were more likely to obtain biological advantage than those who started the process later.

What about incest taboos? Differences between cultures in the close relatives (or presumed close relatives) that are prohibited as targets of sexual interest are not genetically inherited. Many authors have suggested that individuals may derive reproductive success from the taboos.[29] However, those individuals who impose the prohibitions do not derive immediate personal benefits from their behavior. Others derive those direct fitness benefits. Social benefits may be derived because the group does not have to pay the costs of caring for individuals who in various ways are less fit. But note here that an attempt to mount a purely eugenic argument would be confused because the maladaptive genes expressed when inbreeding is common are not removed from the population by preventing inbreeding. Indeed, inbreeding is the best way of getting rid of those genes in the long run. The social benefits of incest taboos may therefore be seen as equivalent to those modern laws that require the wearing of seat belts in cars.

Whether people are aware of the effects of inbreeding is another issue. William Durham has presented persuasive evidence that in many cultures people are aware.[30] He also documents cases where they are not, and we are left with the possibility that either they were aware in the past or incest taboos were driven by other pressures. Awareness of the ill effects of inbreeding would be best translated into the conviction that the aware individual should not have children with his or her sibling. Nothing more is required of Darwinian evolution. The awareness does not immediately translate into a conviction that others should be stopped from having children with their siblings. Or if it does, it is driven by a different utility that involves group selection.

Differences of outcome raise problems for an argument developed by E. O. Wilson. He wrote,

> By translating the Westermarck effect into incest taboos, humans appear to pass from pure instinct to pure rational choice. . . . I suggest that rational choice is the casting about among alternative mental scenarios to hit upon the ones which, in a given context, satisfy the strongest epigenetic rules. It is these rules and this hierarchy of their relative strengths by which human beings have successfully survived and reproduced for hundreds of millennia. The incest avoidance case may illustrate the manner in which the coevolution of genes and culture has woven not just part but all of the rich fabric of human social behavior.[31]

Implicit in this argument is that the incest taboo is serving the same function for the individual as inbreeding avoidance. The same point arises in a formal genes-culture coevolutionary model developed by Aoki and Feldman.[32] Further, the model assumes random mating by those who don't have the postulated "avoid sibling" behavior pattern. Since such random mating does not occur, the assumption renders the model questionable.

In summary, then, I suggest that it is unlikely that inbreeding avoidance and incest taboos evolved by similar mechanisms or even have a common utility in modern life. I fully accept the argument in favor of having both belt and braces (see Chapter 6). Redundant mechanisms are well known in biology, those used in navigation by birds being a famous example. Even so, incest taboos need not necessarily serve the same function as the inhibitions derived from early experience. If that much is accepted, what other mechanism for the cultural evolution of incest taboos should be entertained? Like Wolf, I think that a strong case can be made for the hypothesis advanced by Edward Westermarck.[33]

Westermarck suggested that humans have an inclination to prevent other people from behaving in ways they would not themselves behave. On this view, left-handers were in the past forced to adopt the habits of right-handers because the right-handers found left-handers disturbing. In the same way, those who were known to have had sexual intercourse with close kin were discriminated against. People who had grown up with kin of the opposite sex were generally not attracted to those individuals and disapproved when they discovered others who were. It had nothing to do with society not wanting to look after the half-witted children of inbreeding, since in many cases they had no idea that inbreeding was the cause. Rather, the disapproval was about suppressing abnormal behavior, which is potentially disruptive in small societies.

Such conformity looks harsh to modern eyes, even though we have plenty of examples of it in contemporary life. However, when so much depended on unity of action in the environment in which humans evolved, wayward behavior could have destructive consequences for everybody. It is not difficult to see why conformity should have become a powerful trait in human social behavior. Once in place, the desire for conformity, on the one hand, and the reluctance to inbreed, on the other, would have combined to generate social disapproval of inbreeding. The emergence of incest taboos would take on different forms, depending on which sorts of people, nonkin as well as kin, were likely to be familiar from early life.

If these ideas are correct, human incest taboos did not arise historically from deliberate intention to avoid the biological costs of inbreeding. Rather, in the course of biological evolution, two separate mechanisms appeared. One was a developmental process concerned with striking an optimal balance between inbreeding and outbreeding when choosing a mate. The other was concerned with social conformity. When these two propensities were put together, the result was social disapproval of those who chose partners from within their close family. When social disapproval was combined with language, verbal rules appeared that could be transmitted from generation to generation, first by word of mouth and later in written form.

Conclusion

I believe that the divide between the biological sciences and the social sciences has narrowed to the point where real dialogue occurs. Attempts at a takeover by sociobiologists and, more recently, evolutionary psychologists set this process back, all the more so because they persuaded some social scientists to change their faith and preach with the zeal of the recently converted.

Nevertheless, I believe that we are getting closer. The biologists have come to understand that their own thinking is affected by where they have come from, and the social scientists have started to understand how evidence changes the way they think. Darwinism is no longer the threat it once seemed. It offers explanations that are quite different in character from those provided by studies of how an individual develops. The days of both genetic and environmental determinism are numbered. The center of intellectual activity is now focused on process and how the individual develops.

NOTES

1. P. Bateson, "Optimal outbreeding," in *Mate Choice*, ed. P. Bateson (Cambridge: Cambridge University Press, 1983), pp. 257–77.

2. Ibid.

3. Ibid.

4. Ibid.

5. Ibid.

6. P. Bateson, "Preferences for cousins in Japanese quail," *Nature*, vol. 295 (1982), pp. 236–37.

7. William Durham, *Coevolution: Genes, Culture, and Human Diversity* (Stanford, Calif.: Stanford University Press, 1991).

8. Bateson, "Optimal outbreeding."

9. D. I. Perrett, K. A. May, and S. Yoshikawa, "Facial shape and judgments of female attractiveness," *Nature*, vol. 368 (1994), pp. 239–42.

10. J. Shepher, "Mate selection among second generation kibbutz adolescents and adults: Incest avoidance and negative imprinting," *Archives of Sexual Behavior*, vol. 1 (1971), pp. 293–307.

11. Arthur P. Wolf, *Sexual Attraction and Childhood Association: A Chinese Brief for Edward Westermarck* (Stanford, Calif.: Stanford University Press, 1995).

12. P. Bateson and P. Martin, *Design for a Life: How Behavior and Personality Develop* (New York: Simon and Schuster, 2000).

13. W. Paley, *Natural Theology* (London, Faulder, 1802).

14. Gerald H. Thayer, *Concealing-Coloration in the Animal Kingdom* (New York: MacMillan, 1909).

15. J. S. Gould and R. C. Lewontin, "The spandrels of San Marco and the Panglossian program: A critique of the adaptationist program," *Proceedings of the Royal Society of London, B*, vol. 250 (1979), pp. 281–88.

16. Bateson and Martin, *Design for a Life*.

17. Ibid.

18. J. Shields, *Monozygotic Twins Brought Up Apart and Brought Up Together* (Oxford: Oxford University Press, 1962).

19. Konrad Lorenz, "Der Kumpan in der Umwelt des Vogels," *Journal für Ornithologie*, vol. 83 (1935), pp. 137–213, 289–413.

20. Bateson, "Optimal outbreeding."

21. Shepher, "Mate selection."

22. Wolf, *Sexual Attraction and Childhood Association*.

23. Eleanor E. Maccoby, "Gender and gender relationships," *American Psychologist*, vol. 24 (1990), pp. 523–520.

24. Wolf, *Sexual Attraction and Childhood Association*.

25. P. Bateson, "Rules for changing the rules," in *Evolution from Molecules to Men*, ed. D. S. Bendall (Cambridge: Cambridge University Press, 1983), pp. 483–507.

26. Maccoby, "Gender and gender relationships."

27. Bateson, "Rules for changing the rules."

28. Bateson, "Optimal outbreeding."

29. Durham, *Coevolution*.

30. Ibid.

31. E. O. Wilson, *Consilience: The Unity of Knowledge* (New York: Knopf, 1998).

32. K. Aoki and M. W. Feldman, "A gene-culture coevolutionary model for brother-sister mating," *Proceedings of the National Academy of Sciences, USA*, vol. 94, no. 13 (1997), pp. 46–50.

33. Edward Westermarck, *The History of Human Marriage* (London: Macmillan, 1891).

2 *Genetic Aspects of Inbreeding and Incest*

Alan H. Bittles

When referring to humans, the term *inbreeding* is used to describe unions between couples known to share at least one common ancestor. While now rare in Western societies, marriages between close biological kin are preferential in many parts of the world, including north and sub-Saharan Africa, the Middle East, Central Asia, and much of the Indian subcontinent.[1] Although the rates and types of inbred union may vary according to religious and societal norms, marriages between first cousins are especially common; for example, in Pakistan they currently account for approximately 50 percent of all marital unions.[2]

An incestuous relationship is a union between biological relatives that is genetically closer than permissible under prevailing civil legislation. Most commonly, incest is defined as sexual intercourse between persons defined as first-degree relatives, that is, father-daughter, mother-son, or brother-sister. However, in some countries, such as Scotland, the definition includes half-sib and uncle-niece unions, where the partners have 25 percent of shared genes.[3]

In all forms of consanguineous union the partners share genes inherited from one or more common ancestors; for example, in first-cousin marriages the spouses are predicted to have 12.5 percent of their genes in common. This means that on average their progeny will be homozygous at 6.25 percent of gene loci; that is, they will have inherited identical gene copies from each parent at these sites in their genome. As shown in Table 2.1, this is conventionally expressed as the coefficient of inbreeding (F), which for first-cousin offspring is 0.0625, whereas in matings between first-degree relatives the couple has 50 percent of their genes in common, with an equivalent F value in their progeny of 0.25. Irrespective of the level of inbreeding, if the same mutant gene is inherited from both parents, the individual will express the disorder, prenatally, at birth, or later in life, depend-

TABLE 2.1
Major Types of Consanguineous Relationship

Family Relationship	Genetic Relationship	Fraction of Genes in Common	Coefficient of Inbreeding (F) in Progeny
Incestuous	First degree	1/2	0.25
Uncle-niece	Second degree	1/4	0.125
Double first cousin		1/4	0.125
First cousin	Third degree	1/8	0.0625
First cousin once removed	Fourth degree	1/16	0.0313
Double second cousin		1/16	0.0313
Second cousin	Fifth degree	1/32	0.0156

ing on the nature and site of the mutation, thus contributing to the phenomenon of inbreeding depression.

Attitudes Toward Consanguineous Marriage Within Different Religions

Many examples of consanguineous unions are cited in the biblical texts, with Abraham and Sarah identified as half-brother and sister (Genesis 20: 12), and Amran and Jochebed, the parents of Aaron and Moses, related as nephew and aunt (Exodus 6: 20). At a later date the permitted degrees of marital relationships between biological kin were extensively defined, with marriages up to and including uncle-niece (but not aunt-nephew) permitted (Leviticus 18: 7–18).

In attempting to assess the rationale underpinning these regulations, it is important to acknowledge that many early societies possessed a quite sophisticated knowledge of potentially fatal inherited disorders. For example, the X-linked recessive mode of inheritance of hemophilia A appears to be recognized in the Talmud and other Jewish religious texts dating back to the second to fifth centuries AD. Dispensation for the normally obligatory male circumcision on the eighth day of life could be granted under two sets of circumstances:

1. If a woman had given birth to two or three sons who had died following circumcision, any future male children she might bear would be excused circumcision even if the pregnancy was with a different husband.

2. Where the sons of three sisters had died following circumcision, all male children born to other female siblings would also be excused circumcision.[4]

According to the Venerable Bede, an early ruling on consanguinity within Christianity was given by Pope Gregory I to St. Augustine, the first archbishop of Canterbury, in approximately AD 597.[5] Citing Leviticus 18: 6, the Pope advised that marriages between consanguineous spouses did not result in children and that sacred law forbade a man to "uncover the nakedness of his kindred." For members of the Latin Church, the effect of this ruling was to prohibit marriage with a biological relative, usually up to and including third cousins, although because of different methods of calculating the degrees of biological relationship there was some confusion as to exactly which degrees of consanguineous union were permitted or prohibited. This problem was solved in the canon issued by Pope Alexander II in AD 1076, which resulted in a shift in the formal method of consanguinity classification from the Roman to the Germanic system. For example, third-cousin marriages, which according to the Roman system were of the eighth degree, became fourth degree under the Germanic classification.[6]

The formal proscription of consanguineous unions was confirmed in 1215 by the Fourth Lateran Council and generally remained in force within the Latin church until 1917, when it was limited to unions between couples related as first cousins or closer. Dispensation could, however, be granted at Diocesan level for related couples who wished to marry within the prohibited degrees of consanguinity, albeit with payment of an appropriate benefaction to the church.[7]

Similar strict rules governing consanguineous marriage continue to be applied by the Christian Orthodox Church. Among the Reformed Protestant denominations, the existing biblical guidelines (Leviticus 18: 7–18) were generally adopted, although the closest form of approved union usually has been between first cousins. Somewhat paradoxically, the highest rates of consanguineous unions historically recorded in Europe appear to have been in the southern Roman Catholic countries, where consanguinity was subject to church sanction, rather than the Protestant countries of northern and northwestern Europe.[8]

Among other major world religions—Buddhism, Islam, and Judaism, and in smaller religious communities such as the Zoroastrians/Parsis—attitudes toward close kin marriage are generally favorable or neutral.[9] Cross-cousin marriages also are permitted in the Chinese Taoist/Confucian tradition,[10] and in rural Han communities in the period 1949–67, 0.7 percent to 1.2 percent of marriages were between first cousins.[11] While a generally tolerant approach was adopted toward marriage between biological relatives at the inception of the People's Republic of China,[12] under the terms of the 1981 Marriage Act unions between first cousins became illegal. In the face of this prohibition, it must be assumed that their prevalence has since declined, at least in the majority Han population.

The situation within Hinduism is much more complicated and embodies three sets of regulations, one of which promotes endogamy whereas the other two require exogamy. At the uppermost level of differentiation, all Hindu communities are structured into a number of hierarchical groups, that is, *varnas*, castes, and subcastes, each of which has its own set of rights, duties, and privileges.[13] There are four *varnas*—the Brahmins, Kshatriyas, Vaishyas, and Sudras—and each *varna* is subdivided into castes (*jati*), with caste membership hereditary and basically inviolable. Therefore, with few exceptions, caste endogamy is obligatory. Castes are further subdivided into *gotras*, which are transmitted through the male line and are exogamous insofar as marriages between the children of two brothers are prohibited.[14] In addition, according to *sapinda* regulation, persons related to each other within a certain number of generations on the paternal and maternal sides of a pedigree are forbidden to marry.[15]

The net result is that virtually all Hindu marriages continue to be contracted within caste boundaries, and *gotra* regulations also are observed on a near-universal basis. But at the *sapinda* level, major differences are observed between the peoples of North and those of South India. Among the Aryan Hindus of North India, pedigrees are examined over an average of seven generations for males and five generations for females to ensure avoidance of a consanguineous union.[16] By comparison, uncle-niece marriage and first-cousin unions between a man and his maternal uncle's daughter (mother's brother's daughter) have a long tradition in South India,[17] where they continue to be preferential, especially in the states of Andhra Pradesh, Karnataka, and Tamil Nadu.[18]

The Development of Civil Legislation on Consanguineous Marriage in Western Societies

Prior to the mid-nineteenth century, first-cousin marriages were commonly contracted in many Western societies. From the 1850s onward, an often heated debate was conducted within the scientific and medical communities of North America and Western Europe over the biological effects of close kin marriage.[19] Charles Darwin, who had married his first cousin Emma Wedgwood, became a leading protagonist in the controversy. Through the intervention of Sir John Lubbock, Darwin attempted to persuade the Parliament of Great Britain and Ireland that the prevalence of first-cousin marriages should be determined in the 1871 census. The proposal was denied by a parliamentary select committee, largely on the grounds that the subject was of great sensitivity and any formal inquiry would be unacceptably intrusive. For example, according to the report in Hansard, committee member Gathorne Hardy stated that "he did not see the desirability of holding up families where such marriages had taken place, and the children being anatomised for the benefit of science."[20] Likewise, in the opinion of his colleague Mr. Locke, "This was a piece of the grossest cruelty ever thought of." Darwin's marriage to his cousin had produced eleven children, and he displayed considerable annoyance toward the politicians' rebuttal, observing in *The Descent of Man*, "When the principles of breeding and of inheritance are better understood, we shall not hear ignorant members of our legislature rejecting with scorn a plan for ascertaining by an easy method whether or not consanguineous marriages are injurious to man."[21]

To overcome the resulting stalemate, Darwin's son G. H. Darwin devised a method of assessing the prevalence of first-cousin unions using surname analysis.[22] On presentation of his findings to the Statistical Society in London, the younger Darwin's efforts were praised by Francis Galton, who was half-cousin to Charles Darwin, for the success he had achieved in at least partially correcting "an exaggerated opinion which was current as to the evil resulting from first-cousin marriages."

Within Europe the debate on consanguinity resulted in few if any changes in legislation. However, in the United States the eventual outcome was the passage of laws at state level to control, and in many cases to ban, first-cousin unions, even though much of the information on which these laws were based was of questionable biological validity. Despite a unanimous recommendation by the National Conference of Commissioners on Uniform State Laws that first-cousin marriages be freely permitted in the United States,[23] such marriages remain illegal in twenty-two states and are a criminal offense in eight others, as shown in Figure 2.1.[24] In some states the traditions of specific communities were recognized, and so, for example, uncle-niece marriages are permissible in the Jewish community of Rhode Island, while in Colorado Native Americans may marry their stepchildren.[25] Wisconsin provides an interesting example of a state where legislators had apparently attempted to permit first-cousin unions while circumventing possible adverse biological outcomes; first-cousin marriages are permitted if one or both partners are infertile or the female is over fifty-five years of age and thus presumed to be postmenopausal.[26]

Societies in Which Consanguineous Unions
Are Preferential

As indicated in the introductory paragraph, the current legal situation in the United States differs from that in many other parts of the world.

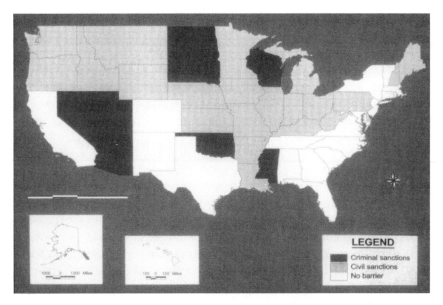

FIGURE 2.1. Map of the United States Indicating States in Which First-Cousin Marriages Are Legal and Those in Which They Are Civil or Criminal Offenses

Preferential consanguineous marriage is mainly explained in social and economic terms, and the reasons given include the strengthening of family relationships and the maintenance of family property, including landholdings. Prenuptial arrangements also are greatly simplified, and the security of marrying a partner whose entire family background is known is perceived as a major benefit in ensuring the success and stability of the union, with lower divorce rates among consanguineous couples.[27] In general terms, the highest rates of consanguineous union are reported in rural areas among the poorest and least educated sections of society, although landowning families also arrange intrafamilial marriages to ensure the maintenance of their landholdings.[28]

Variations in the specific types of marriage contracted—for example, with first-cousin unions between a man and his father's brother's daughter preferred in Arab Muslim communities,[29] as opposed to mother's-brother's-daughter marriage in such disparate populations as the Dravidian Hindu states of South India,[30] the Han Chinese,[31] and the Tuareg of North Africa[32]—indicate additional customary influences. Equally, while uncle-niece unions ($F = 0.125$) are practiced within Judaism, they are proscribed by the Koran, even though Muslims are permitted double first-cousin marriages, which represent the same genetic distance.

The Influence of Consanguinity on Reproductive
Behavior and Success

Based on data collected from the Hutterites, a highly endogamous Anabaptist sect resident in South Dakota, it has been claimed that mate choice is influenced by HLA haplotypes, with a lower-than-expected incidence of HLA-haplotype matches.[33] These studies, which appear to implicate HLA-associated human odor preferences, are preliminary and somewhat controversial, and it has been suggested that the underlying biological mechanism may operate effectively only in communities with very limited HLA repertoires, such as the Hutterites themselves.[34] If this is correct, then similar preconditions could potentially exist in many highly inbred communities. Nevertheless, the relevance of the findings to the majority of present-day consanguineous unions is probably marginal, since in populations where consanguinity is highly preferential, such as the Middle East, South India, and Pakistan, marriages are customarily arranged by the parents of the couple. Thus the potential degree of autonomy in partner choice is very limited.

Enhanced genetic compatibility would be expected between mother and fetus in marriages between close biological kin, resulting in an advantageous pregnancy outcome because of decreased rates of ABO and more especially Rh incompatibility[35] and preeclamptic toxemia.[36] Conversely, according to the fetal allograft concept,[37] antigenic disparity between mother and fetus is beneficial to fetal development.[38] The hypothesis has been extensively examined in human populations, especially by reference to women who have experienced primary recurrent spontaneous abortions with all conceptions lost. In some studies there was a significant positive association between parental allele sharing at HLA loci, principally involving HLA-DR, -DQ, and -B alleles, whereas in others none could be detected.[39]

Even if a positive association does exist between infertility and parental HLA allele sharing, this association is far from complete, as successful pregnancies have been described in which the HLA haplotypes of the mother and offspring were identical.[40] Furthermore, in the highly inbred Hutterite community, couples who shared HLA-DR antigens had a median completed family size of 6.5, as opposed to 9.0 among those with no HLA alleles in common.[41] Accordingly, it has been proposed that the association between parental HLA compatibility and recurrent spontaneous abortion may best be explained in terms of the expression of deleterious recessive genes located in the HLA-DQ-DR-B region of the major histocompatibility complex.[42]

The validity of each of these hypotheses, and their direct relevance to highly inbred unions, has to be judged against the very high spontaneous abortion rates in humans, with on average some 40 percent of all post-implantation conceptions lost.[43] Reduced levels of pathological sterility have been reported in a number of different inbred populations,[44] and other indirect indicators of fetal survival, such as rates of multiple births and the secondary sex ratio, also have failed to show an adverse inbreeding effect.[45]

Conversely, there is preliminary evidence that consanguineous couples enrolled in assisted reproductive technology programs may have higher miscarriage rates,[46] and studies conducted in the Hutterite community indicate somewhat longer birth intervals in the more inbred women, possibly owing either to lower conception rates or higher peri-implantation losses.[47] In addition, DNA-based investigations among highly inbred communities in Pakistan have suggested that there may be selection against homozygosity at specific gene loci involved in embryonic and early fetal development.[48]

Assessed in terms of completed family size, a recent meta-analysis was unable to identify any overall adverse effect of inbreeding on fertility, and in a large majority of the constituent studies consanguineous couples actually had more children.[49] The data did, however, suggest that the greater number of children born to related couples in part reflected reproductive compensation for children dying at a young age.

Estimating the Biological Outcomes of Consanguineous Marriages

As indicated in the introduction, the coefficient of inbreeding (F) is a numerical estimate of the degree of inbreeding of an individual. Of course in many communities there is a long continuous history of consanguineous unions, and so the cumulative level of inbreeding may be significantly higher than the value calculated for a single generation. In such populations a correction can be applied to account for the effects of ancestral inbreeding, using this formula:

$$F = \sum (\tfrac{1}{2})^n (1 + F_A)$$

where F_A is the ancestor's inbreeding coefficient, n is the number of individuals in the path connecting the parents of the individual, and the summation ($\sum$) is taken over each path in the pedigree that goes through a common ancestor.

The first structured study into the biological effects of inbreeding was organized by Samuel Bemiss of Louisville, Kentucky, who collated reports forwarded by medical colleagues on the outcomes of unions ranging from incest to third-cousin marriages.[50] Since that time, many additional studies have been undertaken, based on a variety of sampling techniques including pedigree analysis, household surveys, questionnaires administered to hospital in- and outpatients, Roman Catholic dispensation records, and isonymy (surname analysis). In the populations of Dravidian South India, where uncle-niece and first-cousin unions are preferential and jointly account for some 30 percent of marriages, unions beyond second cousins ($F < 0.0156$) are of very limited biological significance.[51] By comparison, in countries such as Finland, where consanguineous unions generally are rare, biologically remote relationships may result in a moderate level of cumulative inbreeding through time and may be of potential clinical relevance.[52] This situation also applies in endogamous communities in which close cousin marriage has been proscribed on religious grounds but marriages between couples related to a lesser degree are frequently contracted, an example being the Roman Catholic community on the isle of Eriskay in Scotland.[53]

In an attempt to rationalize the results of inbreeding surveys into a meaningful and reproducible format, a method was devised to calculate the numbers of lethal gene equivalents in a community or population by comparing death rates in the progeny of consanguineous and that of unrelated couples. Lethal gene equivalents are the numbers of detrimental recessive genes carried by an individual in the heterozygous state, which, if homozygous, would result in death. In a population, the number of lethal gene equivalents can be calculated according to this formula:

$$-\log_e S = A + BF$$

where S is the proportion of survivors in the study population, A measures all deaths that occur under random mating, B represents all deaths caused by the expression of recessive genes via inbreeding, and F is the coefficient of inbreeding.[54] By plotting a weighted regression of the log proportion of survivors (S) at different levels of inbreeding (F), A can be determined from the intercept on the y-axis at zero inbreeding ($F = 0$), and B (the number of lethal gene equivalents) is given by the slope of the regression. A number of theoretical and methodological limitations have been identified;[55] however, the regression technique offers a simple and convenient method to assess and compare data on the effects of inbreeding in different populations.

Through time estimates of the numbers of lethal gene equivalents per population have been revised downward,[56] in part because of better sampling techniques and the recognition that earlier surveys may have produced spuriously high values owing to inadequate control for nongenetic variables, including socioeconomic status. Other, more direct methods have also been used, and in a study that compared deaths in the offspring of first-cousin and unrelated unions from approximately six months gestation to ten years of age, mortality was 4.4 percent higher in the consanguineous group, equivalent to 1.4 lethal gene equivalents per zygote.[57]

The Phenomenon of Incest Avoidance

With a few notable exceptions, incest avoidance appears to be common to virtually all societies. Even in contemporary fiction—for example, Ian McEwan's *Cement Garden* and A. S. Byatt's *Angels and Insects*—any breach of this accepted norm evokes outrage.[58] Although a genetic rationale cannot be dismissed, the various explanations given for this type of response are mainly social in origin, including Freudian guilt, and theories ascribed to Lévi-Strauss and others that the incest taboo serves both to maintain the existence of the family and to encourage the establishment of affinal relations with other kin groups.

According to the Westermarck hypothesis, incest avoidance can be explained in terms of negative imprinting against close associates of early childhood.[59] Studies conducted on individuals raised under mixed-sex child-rearing regimes in Israeli kibbutzim have been cited in support of the negative imprinting theory, since they rarely marry or enter into sexual relationships.[60] More convincingly, a detailed investigation into the practice of *sim-pua* marriage in Taiwan, whereby a girl is adopted with the expectation that she will marry a son of the adoptive family, clearly showed both lower mean fertility and lesser marital stability.[61] (For further discussion on this topic, see Chapter 4.)

Historical Examples of Dynastic and Nondynastic Incest

There have been a number of societies in which dynastic incest was practiced over multiple generations, including Egypt during the eighteenth and nineteenth dynasties (prior to 332 BC) and the Ptolemaic (323–30 BC) and Roman (30 BC–AD 324) periods, Zoroastrian Iran, the Incas, and the royal families of Hawaii.[62] In Pharaonic Egypt it was believed that the royal bloodline would be maintained and strengthened through brother-sister unions, although the Pharaoh also had ready access to other, unrelated brides and concubines. Increasingly, information has emerged on incestuous marriages outside these ruling classes in which marriage contracts were written and the bride brought her own dowry.[63] For example, in Roman Egypt, full brother-sister unions accounted for 19.6 percent of marriages in the city of Arsinoe, with a further 3.9 percent of marriages between half siblings (see Chapter 5).[64]

The rationale for incestuous unions among Zoroastrians appears to have been based on religious principles and to have been associated with their attitudes toward marriage, which was regarded as divinely favored and almost akin to a religious duty. While unknown in the Avesta, *xvaét-vadatha*, or *xvétxvét*, (usually translated as next-of-kin marriage) was first discussed in the Pahlavi texts of the sixth to ninth centuries AD,[65] with all three types of incestuous union—father-daughter, mother-son, and brother-sister—advocated. *Xvétxvét* was described as being of special religious merit, being the ninth of thirty-nine ways of gaining entry to heaven, and its practice was accepted as a means of expiating mortal sin.

It has been calculated that in early human societies, with restricted numbers of potential marriage partners, an incest taboo could result in considerable demographic costs to the community.[66] The practical desirability of incestuous unions on financial grounds, and to prevent the subdivision of landholdings, was later considered by the Christian Cathar community of Occitania in the Middle Ages,[67] a trait that probably contributed to its suppression by the Latin church. Interestingly, in the light of contemporary information on the adverse health effects of incestuous unions described below, the genealogies of the eighteenth- and nineteenth-dynasty Egyptian kings have given no indication of reduced reproductive capacity, and there was little or no recorded evidence of physical or possible mental defects in the mummies of the royal brother-sister offspring.[68]

The Prevalence and Types of Incestuous Unions in Contemporary Societies

There is the conundrum that, despite the claimed near-universality of incest avoidance, an extensive and growing literature exists on the occurrence of incest in contemporary Western societies. These reports most commonly have emanated from case studies conducted on persons examined because of intellectual handicap,[69] consultations with patients in psychiatric clinics,[70] or individual cases of young children diagnosed with a rare inherited disorder,[71] thus leading to the general conclusion that incestuous relationships are highly detrimental to those classified as victims. There have been occasional reports suggesting that the outcome need not necessarily be psychologically damaging, even when a long-term father-daughter relationship had commenced while the female was still prepubertal.[72] As incest is both illegal and widely regarded as morally reprehensible, it would be difficult to ascertain how commonplace this latter response may be, without exposing individuals to the risk of self-incrimination (for further discussion on this topic, see Chapter 9).

The Biological Outcomes of Incest

Given the strongly judgmental societal attitudes, direct assessment of the biological outcomes of incest may be subject to significant ascertainment bias, as a rigorous examination, including determination of paternity, may be initiated only if a child shows symptoms of physical or intellectual handicap, or both. The net result is a marked lack of data, which probably does not reflect the actual numbers of incestuous pregnancies that successfully proceed to term. Problems also arise in attempting to control for the potentially adverse effects of nongenetic variables, such as very young or advanced maternal and paternal ages, parental disease, and unsuccessful attempted interruption of the pregnancy.

These points are illustrated in Table 2.2, which contains outcome information from the four most comprehensive studies of incest that have been reported to date. In three of the studies the data were collected prospectively,[73] and in the fourth the information was collected from early in postnatal life.[74] In the studies there were equal numbers of father-daughter and sib incest (for each, n = 96), and one mother-son offspring. This is somewhat surprising, as most commentators on the subject have considered father-daughter relationships to be the most common form of incest.

The first and obvious observation is that very few cases of children born to incestuous unions have been systematically investigated, surprisingly so, since the topic features so prominently in the social and behavioral literature. The brief report by C. O. Carter is severely restricted by a lack of information on the physical and mental status of the parents and their socioeconomic background, and by a lack of controls.[75] In one of the studies, strenuous efforts were made to recruit matched nonincestuous controls, and the small numbers are indicative of the difficulties encountered by the investigators in this task.[76]

The numerically largest study was that of E. Seemanová, retrospectively conducted in the former Czechoslovakia on incestuous births reported between 1933 and 1970.[77] The method of control employed is at first sight very convincing, with the outcomes of incestuous matings compared with those of pregnancies undertaken by a subset of the same women with unrelated partners. Unfortunately, the methodology could not overcome the fact that the physical and mental states of this latter group of women, just 33 percent of the original total, was significantly superior to the remaining majority of those studied.[78] For example, of the 141 females in incestuous unions, twenty were intellectually handicapped (of whom two additionally were deaf-mutes, two had congenital syphilis, and two were epileptics), a further two were deaf-mutes and three were schizophrenic. By comparison, among the forty-six control mothers, that is, those who also had pregnan-

TABLE 2.2

Levels of Death and Defect Reported in Four Studies of Incest

Country of Origin	No. Studied	Follow-up (yr.)	Autosomal Recessive Disorders	Congenital Malformations/ Sudden Infant Deaths	Nonspecific Severe Intellectual Handicap	Others, Including Mild Intellectual Handicap	Normal	Source
United States	18	0.5	2	4	0	5	7	Adams and Neel 1967
United Kingdom	13	4–6	2	1	1	4	5	Carter 1967
Czechoslovakia	161	1–37	20	21	24	18	78	Seemanová 1971
Canada	21	0.5–1.9	1	8	0	4	8	Baird and McGillivray 1982
Totals	213		25 (11.7%)	34 (16.0%)	25 (11.7%)	31 (14.6%)	98 (46.0%)	

SOURCE: C. O. Carter, "Risk to offspring of incest," *The Lancet*, vol. 289 (1967), p. 436.

cies with unrelated males, only two were intellectually impaired (one additionally being a deaf-mute) and two others were deaf-mutes. Similarly, of the 138 fathers in the incestuous unions, eight were intellectually handicapped, thirteen were chronic alcoholics, two had syphilis, and four had subsequently committed suicide. Whereas in the nonincestuous control group of fifty-two fathers, none were intellectually handicapped, two were chronic alcoholics, and there was one case of polydactyly.

In such a relatively small-scale study, the actions of a single individual also can greatly influence the overall collated results. According to Seemanová, one alcoholic individual had fathered five children with three different daughters; each of these five children had been diagnosed with varying degrees and types of abnormality, and three had died within the first ten days of life.[79]

The very young age of many of the mothers was a further factor that may have adversely affected the viability and health of the incestuous progeny. The mean and modal maternal ages at time of birth in the father-daughter matings were 18.9 and 16 years, and 19.9 and 14 years in the brother-sister matings, versus 24.9 and 21 years in the married control group. Given the extended time period over which the study was conducted and the marked negative secular trend with regard to menarche in Europe during the middle decades of the twentieth century,[80] it seems probable that many of these females had just entered their reproductive phase when the pregnancy commenced.

Conceptions initiated within two years of menarche may be associated with gynecological immaturity and incomplete pelvic growth, and hence an adverse pregnancy outcome. For example, in the United States pregnancies among black females of fourteen years of age or younger resulted in adverse maternal health outcomes, including acute toxemia, uterine dysfunction, and one-day fever, and their progeny had elevated rates of cardiovascular anomalies.[81] Likewise, the progeny of white females aged thirteen to fifteen years exhibited lower birth weight, prematurity, and small size for gestational age.[82] Some of these differences may have been social in origin and may have reflected the disadvantaged circumstances of the mothers prior to and during pregnancy.[83] However, pregnancies in very young women have been shown to exhibit increased rates of chromosomal anomalies,[84] congenital abnormalities,[85] and neonatal and postneonatal mortality.[86]

With these factors in mind, while unavoidable, the age discrepancies between the mothers in incestuous unions and those in unrelated unions may cause significant difficulties in comparing the test and control groups. Indeed, if the data are censored to exclude physical and mental abnormalities among the male and female parents, and major disparities with respect to young and advanced maternal age, few differences remain in the overall health outcomes recorded for each group.

Combining the cases of specific autosomal recessive disorders, major congenital malformations, and nonspecific severe intellectual handicap, 64 out of 213 (39.4 percent) of the progeny of incestuous unions were adversely affected and 30 (14.1 percent) of these individuals had died. In the two studies for which nonconsanguineous reference groups were available, 9 out of 113 control children died or a serious defect was diagnosed (8.0 percent). Thus the mean excess level of death and severe defect in the offspring of incestuous unions was 31.4 percent, not all of which would necessarily be genetic in origin.

How should these four data sets be assessed? Clearly, the composite results indicate a high rate of physical and mental abnormality. The collection of data on this subject is, however, extremely difficult and, as acknowledged by several of the authors, control for nongenetic variables in the incestuous unions may have been incomplete. Attempted illegal abortions were reported in a number of incestuous pregnancies in the earlier studies, which also may have resulted in nonfatal damage to the developing fetus.[87] Given these various circumstances, the average level of reported death and severe defect (39.4 percent) can perhaps best be considered as an upper bound estimate, but one that may well be inflated by a variety of important nongenetic factors, such as the following:

Young maternal age

Advanced paternal and maternal ages

Physical or mental abnormality, or both, in one or both parents

Low socioeconomic status

Specific adverse environmental and social effects, including
    attempted abortion

The second major method involves calculation of the probable levels of defect in the progeny of incestuous matings, based on information gained from legal consanguineous unions. For example, it is assumed that as the progeny of an incestuous mating have a coefficient of inbreeding of $F = 0.25$, their levels of disease and disability will be approximately four times that observed in first-cousin offspring ($F = 0.0625$). The main advantage of this approach is that data on first-cousin unions are more plentiful, and they are less likely to be subject to ascertainment bias. The potential disadvantage is that the relationship between the degree of inbreeding and rates of disease and disability may be curvilinear in nature, with disproportionate increases in the prevalence of deaths and defects at closer levels of inbreeding. This situation could arise if incest was associated with increased rates of conditional lethals, genes that are expressed only under especially stressful circumstances, as might be encountered in an incestuous pregnancy.

Two lines of evidence suggest that conditional lethals may not form a

American Journal of Human Genetics, vol. 45 (1989), pp. 262–69; Ober et. al, "HLA and mate choice."

35. C. Stern and D. R. Charles, "The Rhesus gene and the effect of consanguinity," *Science*, vol. 101 (1945), pp. 305–7.

36. A. C. Stevenson, B. C. C. Davison, B. Say, S. Ustuoplu, D. Liya, M. Abul-Einem, and H. K. Toppozada, "Contributions of fetal/maternal incompatibility to aetiology of pre-eclamptic toxaemia," *The Lancet*, vol. 298 (1971), pp. 1286–89; A. C. Stevenson, B. Say, S. Ustaoglu, and Z. Durmas, "Aspects of pre-eclamptic toxaemia of pregnancy, consanguinity, and twinning in Ankara," *Journal of Medical Genetics*, vol. 13 (1976), pp. 1–8.

37. B. Clarke and D. R. S. Kirby, "Maintenance of histocompatibility polymorphisms," *Nature*, vol. 211 (1966), 999–1000.

38. M. Adinolfi, "Recurrent habitual abortion, HLA sharing, and deliberate immunization with partner's cells: A controversial topic," *Human Reproduction*, vol. 1 (1986), pp. 45–48; C. Ober, "HLA and pregnancy: The paradox of the fetal allograft," *American Journal of Human Genetics*, vol. 62 (1998), pp. 1–5.

39. A. H. Bittles and P. Matson, "Genetic influences on human fertility," in *Infertility in the Modern World: Present and Future Prospects*, ed. G. R. Bentley and C. G. N. Mascie-Talyor (Cambridge: Cambridge University Press, 2000), pp. 46–81.

40. D. C. Kilpatrick, "A case of materno-fetal histocompatibility—implications for leucocyte transfusion treatment of recurrent abortions," *Scottish Medical Journal*, vol. 29 (1984), 110–12; J. R. Oksenberg, E. Persitz, A. Amat, and C. Brautbar, "Maternal-paternal histocompatibility: Lack of association with habitual abortion," *Fertility and Sterility*, vol. 42 (1984), pp. 389–95.

41. C. Ober, S. Elias, E. O'Brien, D. D. Kostyu, W. W. Hauck, and A. Bombard, "HLA sharing and fertility in Hutterite couples: Evidence for prenatal selection against compatible fetuses," *American Journal of Reproductive Immunology and Microbiology*, vol. 18 (1988), pp. 111–15.

42. K. Jin, H. N. Ho, T. Speed, and T. J. Gill, "Reproductive failure and the major histocompatibility complex," *American Journal of Human Genetics*, vol. 46 (1995), pp. 1456–67.

43. Bittles and Matson, "Genetic influences on human fertility."

44. Y. Yanase, N. Fujiki, Y. Handa, M. Yamaguchi, Y. Kishimato, T. Furusho, et al., "Genetic studies on inbreeding in some Japanese populations. XII. Studies of isolated populations," *Japanese Journal of Human Genetics*, vol. 17 (1973), pp. 332–36; P. S. S. Rao and S. G. Inbaraj, "Inbreeding effects on human reproduction in Tamil Nadu of South India," *Annals of Human Genetics*, vol. 41 (1977), pp. 87–98; M. Edmond and M. De Braekeleer, "Inbreeding effects on fertility and sterility: A case-control study in Saguenay-Lac-Saint-Jean (Quebec, Canada) based on a population registry," *Annals of Human Biology*, vol. 20 (1993), pp. 545–55.

45. A. H. Bittles, A. Radha Rama Devi, and N. Appaji Rao, "Consanguinity, twinning, and secondary sex ratio in the population of Karnataka, South India," *Annals of Human Biology*, vol. 15 (1988), pp. 455–60.

46. P. E. Egbase, M. Al-Sharhan, S. Al-Othman, M. Al-Mutawa, and J. G.

Grudzinskas, "Outcome of assisted reproduction technology in infertile couples of consanguineous marriage," *Journal of Assisted Reproduction and Genetics*, vol. 13 (1996), pp. 279–81.

47. C. Ober, T. Hyslop, and W. W. Hauck, "Inbreeding effects on fertility in humans: Evidence for reproductive compensation," *American Journal of Human Genetics*, vol. 64 (1999), pp. 225–31.

48. W. Wang, S. G. Sullivan, S. Amed, L. A. Zhivotovsky, and A. H. Bittles, "A genome-based study of consanguinity in three co-resident endogamous Pakistan communities," *Annals of Human Genetics*, vol. 64 (2000), pp. 41–49.

49. A. H. Bittles, J. C. Grant, S. G. Sullivan, and R. Hussain, "Does inbreeding lead to decreased human fertility?" *Annals of Human Biology*, vol. 29 (2002), pp. 111–31.

50. S. M. Bemiss, "Report on influence of marriages of consanguinity upon offspring," *Transactions of the American Medical Association*, vol. 11 (1858), 319–425.

51. A. H. Bittles, W. Mason, J. Greene, and N. Appaji Rao, "Reproductive behavior and health in consanguineous marriages," *Science*, vol. 252 (1991), pp. 789–94.

52. E. O'Brien, L. B. Jorde, B. Ronnlof, J. O. Fellman, and A. W. Erickson, "Founder effect and genetic disease in Sottunga, Finland," *American Journal of Physical Anthropology*, vol. 77 (1988), pp. 335–46.

53. A. P. Robinson, "Inbreeding as measured by dispensations and isonymy on a small Hebridean island, Eriskay," *Human Biology*, vol. 55 (1983), pp. 289–95.

54. Newton E. Morton, J. F. Crow, and H. J. Muller, "An estimate of the mutational damage in man from data on consanguineous marriages," *Proceedings of the National Academy of Sciences, USA*, vol. 42 (1956), pp. 855–63.

55. E. Makov and A. H. Bittles, "On the choice of mathematical models for the estimation of lethal gene equivalents in man," *Heredity*, vol. 57 (1986), pp. 377–80.

56. A. H. Bittles and E. Makov, "Inbreeding in human populations: Assessment of the costs," in *Mating Patterns*, ed. C. G. N. Mascie-Taylor and A. J. Boyce (Cambridge: Cambridge University Press, 1988), pp. 153–67.

57. A. H. Bittles and J. V. Neel, "The costs of inbreeding and their implications for variations at the DNA level," *Nature Genetics*, vol. 8 (1994), pp. 117–21.

58. Ian McEwan, *Cement Garden* (London: Jonathan Cape, 1980); A. S. Byatt, *Angels and Insects* (London: Chatto and Windus, 1992).

59. P. L. van den Berghe, "Human inbreeding avoidance: Culture in nature," *Behavioral and Brain Sciences*, vol. 6 (1983), pp. 91–102.

60. Y. Talmon, "The family in a revolutionary movement: The case of the kibbutz in Israel," in *Comparative Family Systems*, ed. M. F. Nimkoff (Boston: Houghton Mifflin, 1965), pp. 259–86; M. E. Spiro, *Children of the Kibbutz* (New York: Schocken Books, 1969), pp. 326–35, 347–50.

61. A. P. Wolf, *Sexual Attraction and Childhood Association: A Chinese Brief for Edward Westermarck* (Stanford, Calif.: Stanford University Press, 1995).

62. R. Middleton, "Brother-sister and father-daughter marriage in Ancient

Egypt," *American Sociological Review*, vol. 27 (1962), pp. 603–11; B. D. Shaw, "Explaining incest: brother-sister marriage in Graeco-Roman Egypt," *Man (N.S.)*, vol. 27 (1992), pp. 267–99.

63. K. Hopkins, "Brother-sister marriage in Roman Egypt," *Comparative Studies in Social History*, vol. 22 (1980), pp. 303–54; W. Scheidel, "Brother-sister and parent-child marriage outside royal families in ancient Egypt and Iran: A challenge to the sociobiological view of incest," *Ethnology and Sociobiology*, vol. 17 (1996), pp. 319–40.

64. W. Scheidel, "Brother-sister marriage in Roman Egypt," *Journal of Biosocial Science*, vol. 29 (1997), 361–71.

65. L. H. Gray, "Iranian marriages," in *Encyclopedia of Religion and Ethics*, vol. 8, ed. J. Hastings (Edinburgh: Clark, 1915), pp. 455–59.

66. E. A. Hammel, C. K. McDaniel, and K. W. Wachter, "Demographic consequences of incest tabus: A microsimulation analysis," *Science*, vol. 205, (1979), pp. 972–77.

67. E. Le Roy Ladurie, *Montaillou: Cathars and Catholics in a French Village, 1294–1324* (London: Penguin, 1980), pp. 36, 52, 179ff.

68. M. A. Ruffer, "On the physical effects of consanguineous marriages in the royal families of Ancient Egypt," in *Studies in the Paleopathology of Egypt*, ed. L. R. Moodie (Chicago: University of Chicago Press, 1922), pp. 322–66.

69. J. Jancar and S. J. Johnston, "Incest and mental handicap," *Journal of Mental Deficiency Research*, vol. 34 (1990), pp. 483–90.

70. M. Virkkunen, "Incest offenders and alchoholism," *Medicine, Science, and the Law*, vol. 14 (1974), pp. 124–28; A. A. Rosenfeld, "Incidence of a history of incest among 18 female psychiatric patients," *American Journal of Psychiatry*, vol. 136 (1979), pp. 191–95.

71. B. H. Bulkley and G. M. Hutchins, "Pompe's disease presenting as hypertrophic myocardiopathy with Wolff-Parkinson-White syndrome," *American Heart Journal*, vol. 96 (1978), pp. 246–52; C. Garrett and H. Tripp, "Unknown syndrome: Mental retardation and postaxial polydactyly, congenital absence of hair, severe seborrhoeic dermatitis, and Perthes' disease of the hip," *Journal of Medical Genetics*, vol. 25 (1988), pp. 270–72; Eva Seemanová, "A study of incestuous matings," *Human Heredity*, vol. 21 (1971), pp. 108–28; J. Delanghe, H. Vlaminck, D. Bernard, E. Robbereacht, M. Paret, M. De Buyzere, et al., "Absence of serum alanine aminotransferase activity," *Clinical Chemistry*, vol. 43 (1997), pp. 1665–67.

72. N. Lukianowicz, "Incest. I Paternal incest. II Other types of incest," *British Journal of Psychiatry*, vol. 120 (1972), pp. 301–13.

73. Morton S. Adams and James V. Neel, "Children of incest," *Pediatrics*, vol. 40 (1967), pp. 55–62; C. O. Carter, "Risk to offspring of incest," *The Lancet*, vol. 289 (1967), p. 436; P. A. Baird and B. McGillivray, "Children of incest," *Journal of Pediatrics*, vol. 101 (1982), pp. 854–57.

74. Seemanová, "A study of incestuous matings."

75. Carter, "Risk to offspring."

76. Adams and Neel, "Children of incest."

77. Seemanová, "A study of incestuous matings."

78. A. H. Bittles, "Incest re-assessed," *Nature*, vol. 280 (1979), p. 107.

79. Seemanová, "A study of incestuous matings."

80. J. M. Tanner, *Growth at Adolescence* (Oxford: Blackwell, 1962).

81. J. B. Coates, "Obstetrics in the very young adolescent," *American Journal of Obstetrics and Gynecology*, vol. 108 (1970), pp. 68–72.

82. A. M. Fraser, J. E. Brockert, and R. H. Ward, "Association of young maternal age with adverse reproductive outcomes," *New England Journal of Medicine*, vol. 332 (1995), pp. 113–17.

83. R. L. Goldenberg and L. V. Klerman, "Adolescent pregnancy—another look," *New England Journal of Medicine*, vol. 332 (1995), pp. 1161–62.

84. D. H. Carr, "Chromosomes and abortion," *Advances in Human Genetics*, vol. 2 (1971), pp. 210–57.

85. C. H. Hendricks, "Congenital malformations: Analysis of the 1953 Ohio records," *Obstetrics and Gynecology*, vol. 6 (1955), pp. 592–98.

86. J. Yerushalmy, "Neonatal mortality by order of birth and age of parents," *American Journal of Hygiene*, vol. 28 (1938), pp. 244–70; R. R. Gordon and R. Sunderland, "Maternal age, illegitimacy, and postneonatal mortality," *British Medical Journal*, 297 (1988), p. 774.

87. Adams and Neel, "Children of incest;" Seemanová, "A study of incestuous matings."

88. Bittles and Neel, "The costs of inbreeding."

89. Ibid.

90. William J. Schull and James V. Neel, *The Effects of Inbreeding on Japanese Children* (New York: Harper and Row, 1965).

91. Ibid.

92. A. H. Bittles, "Consanguinity and its relevance to clinical genetics," *Clinical Genetics*, vol. 60 (2001), pp. 89–98; A. H. Bittles, "Consanguineous marriages and childhood health," *Developmental Medicine*, vol. 45 (2003), pp. 571–76.

93. J. G. Hall, "Children of incest: When to suspect and how to evaluate," *American Journal of Diseases of Children*, vol. 132 (1978), p. 1045 (letter); Baird and McGillivray, "Children of incest."

94. J. H. Edwards, "Evidence of incest based on homozygosity," *Annals of Human Genetics*, vol. 52 (1988), pp. 351–53; R. E. Wenk, F. A. Chiafari, and T. Houtz, "Incest diagnosis by comparison of alleles of mother and offspring at highly heterozygous loci," *Transfusion*, vol. 34 (1994), pp. 172–75.

95. R. A. Wells, B. Wonke, and S. L. Thein, "Prediction of consanguinity using human DNA fingerprints," *Journal of Medical Genetics*, vol. 25 (1988), pp. 660–62.

96. S. Bundey, "The child of an incestuous union," in *Medical Aspects of Adoption and Foster Care*, ed. S. Wolkind (London: Spastics International Medical Publications, 1979), pp. 36–41.

97. Baird and McGillivray, "Children of incest."

alternative hypothesis that competition completely accounts for these dispersal patterns,[14] immigrants in several species are not driven out of their natal group and actually receive more aggression in the new group.[15]

Avoidance of Sexual Behavior Between Close Relatives Residing Together

Even though dispersal greatly reduces the chances of close inbreeding, there are sometimes circumstances in which close relatives do remain for at least a time in proximity as adults. Male baboons start mating and are often fertile before they leave their natal troop, and in provisioned and captive groups of macaques, males sometimes remain in their natal group as adults. Also, in chimpanzees, some females are still alive when their sons become mature, and in both gorillas and chimpanzees, young females sometimes start mating in their natal group when their fathers are still alive and adult brothers are present. We now know from long-term studies that although these individuals are sexually active, they often avoid mating with relatives. Table 3.1 lists long-term studies where pedigrees are known and in which the frequency of mating between close relatives has been investigated. In most species in which adult sons are observed with maternal relatives, sexual activity between these males and their mothers is negligible, and it is also greatly reduced between maternal siblings. Significant reduction of sexual activity has also been detected among individuals as distantly related as maternal first cousins in some species where these relationships are known. In some cases—ring-tailed lemurs,[16] Barbary macaques,[17] and rhesus monkeys[18]—paternity analysis has confirmed that matrilineally inbred offspring are seldom produced.

Avoidance of sex with paternal relatives is much less well studied, mostly because in most primate species females mate with several males in the group, so paternal relationships cannot be determined by observational data alone. However, in gorillas, the silverback is often the only breeding male, and very little mating is seen between fathers and daughters. In baboons, Craig Packer found that sexual activity was reduced between females and males that had been present in the troop when the female was born (and were thus potential fathers),[19] and in chimpanzees some females avoided males old enough to be their fathers,[20] while others did not.[21]

Recently, analysis of DNA has allowed precise identification of paternal relatives in some species. A detailed study of a large population of captive Barbary macaques living in an outdoor enclosure found no inhibition of mating between paternal siblings or daughters and their fathers compared to mating with unrelated individuals.[22] Similarly, in a captive group of rhesus

TABLE 3.1
Avoidance of Sexual Activity with Relatives

Species	Mother	Maternal Siblings	Other Maternal Relatives	Father	Paternal Siblings
Ring-tailed lemurs[1]	+	+			−
Muriquis[2]	+				
Marmosets[3]	+			+	
Vervet monkeys[4]	+	+	+		
Japanese macaques[5]	+	+	+		
Rhesus macaques[6]	+	+	+	+	−
Barbary macaques[7]	+	+	+	−	−
Stumptail macaques[8]	+	+			
Olive baboons[9]	+	+		+	
Yellow baboons[10]	+	+	+		+
Chimpanzees[11]	+	+		+/-	
Gorillas[12]				+	

NOTE: + signifies that inhibition of mating occurs, - that it does not; +/- means that inhibition occurs between some pairs but not others. Cells remain blank if the frequency of mating in this category has not been measured in this species.

SOURCES:

[1] M. E. Pereira and M. L. Weiss, "Female mate choice, male migration, and the threat of infanticide in ringtailed lemurs," *Behavioral Ecology and Sociobiology*, vol. 28 (1991), pp. 141–52

[2] K. B. Strier, "Mate preferences of wild muriqui monkeys (*Barchyteles arachnoides*): Reproductive and social correlates," *Folia Primatologica*, vol. 68 (1997), pp. 120–33.

[3] D. H. Abbott, "Behavioral and physiological suppression of fertility in subordinate marmoset monkeys," *American Journal of Primatology*, vol. 6 (1984), pp. 169–86; J. V. Baker, D. H. Abbott, and W. Saltzman, "Social determinants of reproductive failure in male common marmosets housed with their natal family, *Animal Behaviour*, vol. 58 (1999), pp. 501–13; J. A. French, "Proximate regulation of singular breeding in Callitricid primates," in *Cooperative Breeding in Mammals*, ed. N. G. Solomon and J. F. French (Cambridge: Cambridge University Press, 1997), pp. 34–75.

[4] C. A. Bramblett, "Incest avoidance in socially living veret monkeys," *American Journal of Primatology*, vol. 63 (1983), p. 176; Dorothy L. Cheney and Robert M. Seyfarth, personal communication.

[5] K. Tokuda, "A study on the sexual behavior in the Japanese monkey troop," *Primates*, vol. 3 (1961–62), pp. 1–40; T. Enomoto, "The sexual behavior of Japanese monkeys," *Journal of Human Evolution*, vol. 3 (1974), pp. 351–72; Y. Takahata, "The socio-sexual behavior of Japanese monkeys," *Zeitschrift fur Tierpsychogie*, vol. 59 (1982), pp. 89–108.

[6] D. S. Sade, "Inhibition of mother-son mating among free-ranging rhesus monkeys," *Science and Psychoanalysis*, vol. 12 (1968), pp. 18–38; D. G. Smith, "Inbreeding in three captive groups of rhesus macaques," *American Journal of Physical Anthropology*, vol. 58 (1982), pp. 447–51; D. G. Smith, "Avoidance of close consanguineous inbreeding in captive groups of rhesus macaques," *American Journal of Primatology*, vol. 35 (1995), pp. 31–40; B. Chapais, "Male dominance and reproductive activity in rhesus monkeys," in *Primate Social Relationships*, ed. R. A. Hinde (Oxford: Blackwell, 1983), pp. 267–71; J. H. Manson and S. E. Perry, "Inbreeding avoidance in rhesus macaques: Whose choice? *American Journal of Physical Anthropology*, vol. 90 (1993), pp. 335–44.

[7] A. Paul and J. Kuester, "Intergroup transfer and incest avoidance in semi-free-ranging Barbary macaques (*Macaca sylvanus*) at Salem (FRG)," *American Journal of Primatology*, vol. 8 (1985), pp. 317–22; J. Kuester, A. Paul, and J. Arnemann, "Kinship, familiarity, and mating avoidance in Barbary macaques, *Macaca sylvanus*," *Animal Behaviour*, vol. 48 (1994), pp. 1183–94.

[8] R. Daniel Murray and E. O. Smith, "The role of dominance and intrafamilial bonding in the avoidance of close inbreeding," *Journal of Human Evolution*, vol. 12 (1983), pp. 481–86.

[9] Craig Packer, "Inter-troop transfer and inbreeding avoidance in *Papio anubis*," *Animal Behaviour*, vol. 27 (1979), pp. 1–36; L. M. Scott, "Reproductive behavior in adolescent female baboons (*Papio anubis*)," in *Female Primates: Studies by Women Primatologists*, ed. M. Small (New York: Alan Liss, 1984), pp. 77–100; B. B. Smuts, *Sex and Friendship in Baboons* (New York: Aldine, 1985).

[10] S. C. Alberts and J. Altmann, "Balancing costs and opportunities: Dispersal in male baboons," *American Naturalist*, vol. 145 (1995), pp. 179–306; S. C. Alberts, "Paternal kin discrimination in wild baboons," *Proceedings of the Royal Society, London, Series B*, vol. 266 (1999), pp. 1501–6.

[11] C. E. G. Tutin, "Sexual behaviour and mating patterns in a community of wild chimpanzees (*Pan troglodytes schweinfurthii*). Ph.D. thesis, 1975. Zoology. Edinburgh, University of Edinburgh; A. E. Pusey, "Inbreeding avoidance in chimpanzees," *Animal Behaviour*, vol. 28 (1980), p. 543; J. Goodall, *The Chimpanzees of Gombe* (Cambridge: Harvard University Press, 1986).

[12] K. J. Stewart and A. H. Harcourt, "Gorillas: Variation in female relationships," in *Primate Societies*, ed. B. B. Smuts, D. L. Cheney, R. M. Seyfarth, T. T. Struhsaker, and R. W. Wrangham (Chicago: University of Chicago Press, 1987), pp. 358–69.

macaques, David G. Smith found that, in general, inbreeding between patrilineal relatives was not different from expected were mating random.[23] However, father-daughter inbreeding was less frequent than expected. In yellow baboons, Susan Alberts found that although consorting between paternal siblings occurred as often as between nonkin, levels of sexual activity and affiliative behavior were lower during consortships between paternal siblings.[24] Finally, in chimpanzees, two females mated as often as expected with their fathers, while one avoided her father at high rates.[25]

Mechanisms of Kin Recognition

Biologists have proposed two ways that individuals might recognize kin.[26] First, they may use phenotype matching, in which the individual matches cues from other individuals, such as odor, to cues from themselves or their relatives and uses the degree of similarity to determine the degree of kinship. This mechanism has been demonstrated in rodents. For example, inbred mice that differ only in genes at the major histocompatibility complex (MHC) prefer to mate with individuals that differ from themselves in MHC haplotype. MHC genes influence odors in the urine, and this is the cue that the mice match.[27] Cross-fostering experiments show that mice actually match the odor to odors of their parents and choose mates that differ from their parents. Evidence of phenotype matching has also been found in several other species of rodents, crickets, and tadpoles.[28]

The second mechanism, often labeled association or familiarity, is when individuals treat as kin others with whom they had a close association in immaturity. This is the mechanism that Westermarck first proposed to explain incest avoidance.[29] Association has been shown to be important in mate choice in a variety of species of mammals, and experiments in which degree of familiarity is manipulated show that it often has a stronger effect than real kinship.[30] For example, cross-fostering experiments in various species of rodents in which relatives are raised apart and nonrelatives are raised together show that individuals raised apart prefer each other as mates over those reared together, regardless of relatedness.[31]

Among primates, the patterns of inbreeding avoidance described above are consistent with familiarity being the primary mechanism of kin recognition. In all primates, mother-offspring bonds are close and long-lasting, with the result that maternal siblings are often reared in close proximity to one another. In species that show female philopatry, females usually have closer and more affiliative relationships with their close adult female kin. Therefore the fact that mating is strongly inhibited among all kinds of maternal kin in several species is consistent with the Westermarck effect.[32]

In contrast to the consistency of the mother-offspring bond across primate species, close bonds between fathers and daughters occur only in some species. In monogamous species like marmosets, or species like gorillas in which one male has exclusive reproductive access to group females for periods beyond their daughters' age of maturity, father-daughter bonds are strong. In these, father-daughter mating is strongly inhibited (Table 3.1). In other species such as baboons and wild macaques, males do not often remain in the group until their daughters mature, and when they do, there can be great variation in the strength of their relationships with their potential offspring in their first few years of life.

In Barbary macaques, males form close caretaking relationships with several infants in the troop, during which they carry them around and protect them. On the basis of the Westermarck effect, we might expect females that were cared for by some males and not others to avoid males that cared for them regardless of kinship, and this is indeed what is observed in Barbary macaques.[33] While there was no evidence that daughters avoided their fathers (as determined from DNA), mating between females and their previous caretakers was inhibited, and the strength of the inhibition depended on the length of the caretaking relationship. Packer's finding, that females were more likely to avoid consorting and mating with males that had been present at the time of their birth than males that arrived in the troop after their birth, is also consistent with the Westermarck effect.[34] In chimpanzees, females usually leave their natal community, but some stay. Chimpanzee males do not obviously form close relationships with particular infants, although they are frequently tolerant and friendly to most infants.[35] At Gombe, of three females that mated in their community while their fathers were still alive, two did not obviously avoid mating with their fathers, but one did.[36] Future research should determine whether there were differences in early association between the male and female in these cases.

Data on paternal siblings are even scarcer than on fathers and daughters. While the studies of Barbary macaques and rhesus macaques found no significant avoidance of sexual activity between patrilineal siblings,[37] Alberts's baboon study found subtle differences, some of which are consistent with the Westermarck effect.[38] Alberts compared paternal siblings and nonkin on two measures: (1) the rates of consorting activity, when a male maintains a close, guarding relationship with a female during her period of receptivity, and (2) the level of cohesiveness including sexual behavior within consorting pairs. She found that, overall, paternal siblings consorted as often as nonkin did but had lower levels of cohesiveness. She reasoned that one way that paternal siblings might recognize each other is through association. In this population, because the alpha male fathers most of the offspring conceived during his tenure, age cohorts are likely to be paternal kin. Age cohorts also spend more time together than individuals born more than two years apart. Alberts found that individuals within the same age cohort consorted at lower rates than others, regardless of kinship, and also showed lower cohesiveness. She found that on the measure of consortship cohesiveness the three pairs scoring lowest were from the same age cohort and that one pair of baboons that was unrelated but born within two years of each other scored the lowest among unrelated pairs and the highest among age-cohort members. She concluded on the basis of this and the fact that cohesiveness is low between paternal siblings that are not in the same age cohort, that although association is clearly important, phenotype matching may also play a role in the recognition of paternal siblings.

Although systematic rearing experiments have not been done in primates to distinguish between the importance of relatedness per se and familiarity in mating avoidance, some anecdotal evidence supports the importance of the Westermarck effect. Unrelated chimpanzees raised together mated very little with each other and preferred unfamiliar individuals as mates.[39] Also, a male Japanese macaque avoided mating with an unrelated female who had been a close associate of his mother, although he mated readily with unfamiliar females of the same age.[40]

The Ontogeny of Avoidance of Sexual Activity Between Relatives

In many primate species, elements of sexual behavior appear during immaturity. Therefore it is of interest to examine the extent to which such behavior is directed toward relatives compared to nonrelatives and when inhibition between relatives first appears. Data are still very incomplete on this point. Young male macaques,[41] baboons,[42] and chimpanzees[43] get penile erections in the first year of life, and as infants and juveniles they often thrust on other immature individuals during play or on others including adults during play and other social contact. In species in which females exhibit sexual swellings, immature males sometimes mount, achieve intromission, and thrust on swollen, semiswollen, and sometimes unswollen females in a similar way to adult males several years before they are capable of ejaculation. Although immature females in these species may be mounted and thrust upon by immature males, no intromission appears to occur until the females reach adolescence and exhibit sexual swellings or other forms of sexual receptivity.

How much of this sexual behavior by immature males is directed at relatives? Immature sons have been reported to copulate with their mothers in several species. One study found that among captive Japanese macaques, infant males copulated with their mothers, but that juvenile and adolescent males rarely did so and never achieved ejaculation.[44] Male rhesus monkeys reach puberty at the age of about four years.[45] One study found that most males aged three to five years showed no copulatory behavior, but those that did mated almost exclusively with their mothers and maternal sisters.[46] However, they did not show the typical consorting behavior in which a male follows the female, grooms her, and has repeated mount series. Males older than five years that stayed in their natal group mated with other females, but not with their mother or sisters. Two infant male baboons were seen to copulate with their mothers,[47] and a juvenile male was observed to copulate several times with his cycling sister, but mating between postpubertal males and their female relatives was very rare.[48] In Packer's detailed study of the mating behavior of postpubertal males, three males had mothers or sisters that were sexually receptive but never consorted them.[49] One male achieved a complete copulation with his pregnant mother, and on another occasion mounted his mother without intromission when her sexual swelling had just deflated. Another male mounted his lactating mother. In these cases the mother made aggressive vocalizations and terminated the mount.

The most detailed information on the incidence of mating with relatives by immature males comes from chimpanzees.[50] Infant males as young as three years old are able to copulate and thrust with intromission on swollen females when these present their swellings to them, but the males do not reach puberty for many years after this. Mothers in the chimpanzees of Gombe National Park, Tanzania, generally do not resume sexual cycles until their infants are three to six years old, although one or two have resumed cycles within the first year of their infant's life. Males in this population are not capable of ejaculation until they are nine to ten years old. A recent study found that while males of one to two years did not copulate with their swollen mothers, males of three to six years usually copulated quite frequently with their mothers, accounting for about 5 to 7 percent (range 0 to 29 percent) of their mother's total copulations with males when she was swollen (Fig 3.1a).[51] One context in which such copulations occurred was when mothers presented to their distressed infants during weaning.[52] However, many copulations with the mother by males of this age occurred when both male and mother were calm.[53] From the age of seven years onward, most males never copulated with their mothers (Fig 3.1a). However, two of five adult males, Goblin and Frodo, did mate with their mothers several times after the ages of fifteen when they had attained high rank among the males. In these cases the mother usually screamed and strenuously tried to resist her son's advances.[54] In a captive group, a five-year-old juvenile male frequently copulated with swollen females, showing typical male courtship to them beforehand.[55] He also occasionally copulated with his mother, even though she had not yet resumed sexual cycles.

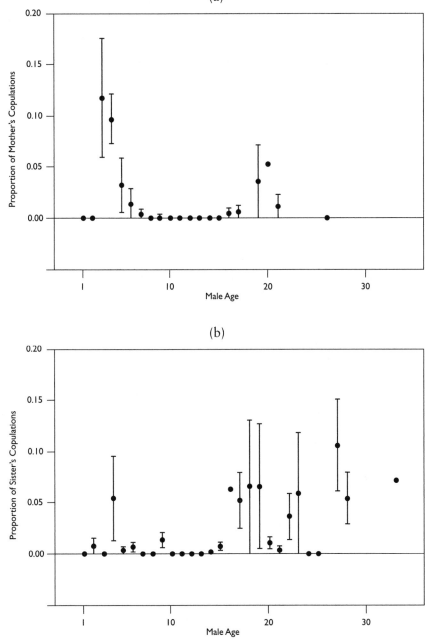

FIGURE 3.1. Copulations Between Mothers and Sons and Sisters and Brothers in the Gombe Chimpanzees. *(a)* Mean (SE) proportion of mothers' copulations contributed by sons of each age (N = 25 mother-son pairs, with some pairs observed over several years). *(b)* Mean (SE) proportion of sisters' copulations contributed by brothers of each age (N = 22 sister-brother pairs, with some pairs observed over several years). (Data from Pusey, Schumacher Stankey, and Goodall, in preparation.)

These copulations generally took place when he appeared tired and depressed, and they were unusual in form. He showed no courtship, the mother did not stand or crouch for her son, and his intromission and thrusting were much more slowly paced and of longer duration than usual. If she prevented these attempts by covering her genitalia, the male "hoo whimpered" and even threw temper tantrums.[56]

At Gombe, males of three to four years mated with their mature sisters, accounting for about 4 percent of their sisters' total copulations, but between the ages of five and fifteen years, males rarely mated with their sisters (Fig 3.1b). In those cases in which females remained in their natal group as adults, some brothers aged sixteen to thirty-three years mated with their sisters at quite high rates, although the sisters often resisted, while other brothers never did so (Fig 3.1b).[57]

Taken as a whole, the data from all these species suggest that males mate with their mothers and, sometimes, maternal sisters with impunity as infants, but that inhibition of such activity sets in before or at puberty. More detailed observations are required to determine whether this inhibition is due to an intrinsic change in the male or is triggered by an increase in resistance from the female. Jane Goodall described how one female chimpanzee allowed her three-year-old infant brother to copulate while she was getting adolescent swellings but prevented him from doing so once she began mating with adult males.[58] However, he was observed copulating with her several times after this.[59]

As well as sexual behavior, infant and juvenile primates exhibit most other patterns of adult behavior, including aggressive behavior. Although these patterns are sometimes performed in an adultlike context, they are often performed during play.[60] Perhaps the sexual behavior of immature individuals is best regarded as play and practice, and the tolerance of female relatives during sexual and other social interactions with these males as investment in improving the social skills of their kin. In all species, copulation among maternal relatives virtually ceases before the male is fully fertile, so the female is at little risk of inbreeding.

The Attachment Theory

How do the primate data speak to the idea that it is attachment and caretaking relationships that preclude sexual behavior in adulthood, rather than other kinds of familiarity? Certainly the relationship of offspring and their mothers is one of attachment and caretaking. This is also likely to be true of relationships between infants and their older siblings. In most primates, females show many caretaking behaviors toward their young siblings, and in at least some, males do so too. In gorillas and some species living in multimale groups, males direct caretaking behavior to infants, and sexual behavior is least common among pairs that were involved in such relationships.

The inhibition of sexual behavior between age cohort mates in baboons is less easily explained by this idea.[61] These individuals are close to the same age and are likely to participate in playful rather than caretaking relationships during immaturity. Finally, the unusual case of hamadryas baboons raises some problems for the idea that caretaking behavior precludes later sexual behavior. Hamadryas baboons form one-male units within larger bands in which an adult male is closely bonded to several adult females with whom he mates.[62] One way in which new units form is for a young adult male to coax an infant female away from her mother and eventually keep her as a mate.[63] During this gradual process the male carries and cares for the infant. At first he directs no sexual behavior toward her, but later he may attempt to mate. Females in this species have sexual swellings, and males have been observed to try to mate with females before they have started swelling. In several cases, such females have produced precocious sexual swellings at earlier ages than females that remain with their natal unit, and begin sexual behavior earlier. The only way to reconcile these observations with the theory that it is attachment in immaturity that precludes sexual behavior later on would be to propose that the sensitive period for this to be effective occurs very early in infancy and is already past for the female by the time a male starts to adopt her.

Conclusion

Nonhuman primates provide abundant evidence for an inhibition of sexual behavior among closely related adults. This finding is consistent with the idea that inbreeding avoidance behavior is a naturally selected behavior that was already present among animals before humans evolved. The primate data support Westermarck's theory that familiarity during immaturity is a major reason for this avoidance. Mating among adults is most inhibited among maternal relatives in species in which these have close associations. The extent to which mating is inhibited among close paternal relatives is more variable and appears to depend largely, though perhaps not completely, on the closeness of association during immaturity. It is also becoming clear that extensive sexual behavior by immature males with close female relatives sometimes occurs before sexual maturity in several nonhuman primate species, but it stops before the risk of conception. More study is required to understand the proximate causes for the waning of this behavior. The primate data are less clear about the kind of familiarity in immaturity that is necessary to prevent mating among adults. Although some relationships, such as those between mother and son, and between older and younger siblings, fit the idea that attachment and caretaking relationships are important, inhibition of sexual behavior between peers of the same age in baboons does not.

NOTES

1. Pierre L. van den Berghe, "Human inbreeding avoidance," *Behavioral and Brain Sciences*, vol. 6 (1982), pp. 91–124.

2. Edward Westermarck, *The History of Human Marriage* (London: Macmillan, 1891).

3. Anne Pusey and Marisa Wolf, "Inbreeding avoidance in animals, *Trends in Ecology and Evolution*, vol. 11 (1996), pp. 201–6.

4. K. Ralls, J. D. Ballou, and A. Templeton, "Estimates of lethal equivalents and the costs of inbreeding in mammals," *Conservation Biology*, vol. 2 (1988), pp. 185–93.

5. Pusey and Wolf, "Inbreeding avoidance in animals."

6. Anne Pusey, "Mechanisms of inbreeding avoidance in nonhuman primates," in *Pedophilia: Biosocial Dimensions*, ed. J. R. Feierman (New York: Springer-Verlag, 1990), pp. 201–20.

7. Craig Packer, D. A. Collins, A. Sindimwo, and Jane Goodall, "Reproductive constraints on aggressive competition in female baboons," *Nature*, vol. 373 (1995), pp. 60–63.

8. Anne Pusey and Craig Packer, "Dispersal and philopatry," in *Primate Societies*, ed. B. B. Smuts, D. L. Cheney, R. M. Seyfarth, T. T. Struhsaker, and R. W. Wrangham (Chicago: University of Chicago Press, 1987), pp. 250–66.

9. M. M. Symington, "Sex ratio and maternal rank in wild spider monkeys: When daughters disperse," *Behavioral Ecology and Sociobiology*, vol. 20 (1987), pp. 333–35.

10. K. B. Strier, "Mate preferences of wild muriqui monkeys (*Barchyteles arachnoides*): Reproductive and social correlates," *Folia Primatologica*, vol. 68 (1997), pp. 120–33.

11. Pusey and Packer, "Dispersal and philopatry."

12. Paul J. Greenwood, "Mating systems, philopatry, and dispersal in birds and mammals," *Animal Behaviour*, vol. 28 (1980), pp. 1140–62.

13. Pusey and Packer, "Dispersal and philopatry."

14. Jim Moore and Rauf Ali, "Are dispersal and inbreeding related?" *Animal Behaviour*, vol. 32 (1984), pp. 94–112.

15. Craig Packer, "Dispersal and inbreeding avoidance," *Animal Behaviour*, vol. 33 (1985), pp. 676–78; Anne Pusey, "Sex-biased dispersal and inbreeding

avoidance in birds and mammals," *Trends in Ecology and Evolution*, vol. 2 (1987), pp. 295–99.

16. M. E. Pereira and M. L. Weiss, "Female mate choice, male migration, and the threat of infanticide in ringtailed lemurs," *Behavioral Ecology and Sociobiology*, vol. 28 (1991), pp. 141–52.

17. Jutta Kuester, Andreas Paul, and J. Arnemann, "Kinship, familiarity, and mating avoidance in Barbary macaques, *Macaca sylvanus*," *Animal Behaviour*, vol. 48 (1994), pp. 1183–94.

18. David G. Smith, "Avoidance of close consanguineous inbreeding in captive groups of rhesus monkeys," *American Journal of Primatology*, vol. 35 (1995), pp. 31–40.

19. Craig Packer, "Inter-troop transfer and inbreeding avoidance in *Papio anubis*," *Animal Behaviour*, vol. 27 (1979), pp. 1–36.

20. Caroline E. G. Tutin, "Sexual behaviour and mating patterns in a community of wild chimpanzees (*Pan troglodytes schweinfurthii*). Ph.D. thesis, 1975. Zoology. Edinburgh, University of Edinburgh; Anne Pusey, "Inbreeding avoidance in chimpanzees," *Animal Behaviour*, vol. 28 (1980), p. 543.

21. Jane Goodall, *The Chimpanzees of Gombe* (Cambridge: Harvard University Press, 1986).

22. Kuester, Paul, and Arnemann, "Kinship, familiarity, and mating avoidance."

23. Smith, "Avoidance of close consanguineous inbreeding."

24. Susan Alberts, "Paternal kin discrimination in wild baboons," *Proceedings of the Royal Society, London, Series B.*, vol. 266 (1999), pp. 1501–6.

25. Anne Pusey, J. Schumacher Stankey, and Jane Goodall, "Age changes in frequency of mating between consanguineous chimpanzees," in preparation.

26. W. G. Holmes and P. W. Sherman, "Kin recognition in animals," *American Scientist*, vol. 71 (1983), pp. 46–55.

27. J. L. Brown and A. Eklund, "Kin recognition and the major histocompatibility complex: An integrative review," *American Naturalist*, vol. 143 (1994), pp. 435–61.

28. David J. C. Fletcher and Charles D. Michener, eds., *Kin Recognition in Animals* (New York: Wiley, 1987); P. G. Hepper, *Kin Recognition* (Cambridge: Cambridge University Press, 1991); P. W. Sherman, H. K. Reeve, and D. W. Pfennig, "Recognition systems," in *Behavioural Ecology*, ed. J. R. Krebs and N. B. Davies (Oxford: Blackwell, 1997).

29. Westermarck, *The History of Human Marriage*.

30. Pusey and Wolf, "Inbreeding avoidance in animals."

31. D. A. Dewsbury, "Kin discrimination and reproductive behavior in muroid rodents," *Behavior Genetics*, vol. 18 (1988), pp. 525–36.

32. Jeffrey R. Walters, "Kin recognition in nonhuman primates," in *Kin Recognition in Animals*, ed. David J. C. Fletcher and Charles D. Michener (New York: Wiley, 1987), pp. 359–93.

33. Kuester, Paul, and Arnemann, "Kinship, familiarity, and mating avoidance."

34. Packer, "Inter-troop transfer and inbreeding avoidance."

35. Jane Goodall, *The Chimpanzees of Gombe*.

36. Pusey, Schumacher Stankey, and Goodall, "Age changes in frequency of mating."

37. Kuester, Paul, and Arnemann, "Kinship, familiarity, and mating avoidance"; Smith, "Avoidance of close consanguineous inbreeding."

38. Alberts, "Paternal kin discrimination in wild baboons."

39. C. L. Coe, A. C. Connolly, H. C. Kremer, and S. Levine, "Reproductive development and behavior of captive female chimpanzees," *Primates*, vol. 20 (1979), pp. 571–82.

40. N. Itoigawa, K. Negamaya, and K. Kondo, "Experimental study of sexual behavior between mother and son in Japanese monkeys," *Primates*, vol. 22 (1981), pp. 494–502.

41. J. P. Hanby and C. E. Brown, "The development of sociosexual behaviors in Japanese macaques, *Macaca fuscata*," *Behaviour*, vol. 49 (1974), pp. 152–96.

42. L. T. Nash, "The development of the mother-infant relationship in wild baboons (*Papio anubis*)," *Animal Behaviour*, vol. 26 (1978), pp. 749–59.

43. Jane Goodall, "The behaviour of free-living chimpanzees of the Gombe Stream Reserve," *Animal Behavior Monographs*, vol. 1, pp. 161–311; Jane Goodall, *The Chimpanzees of Gombe*; Anne Pusey, "Behavioural changes at adolescence in chimpanzees," *Behaviour*, vol. 115 (1990), p. 203; E. S. Savage and C. Malik, "Play and socio-sexual behaviour in a captive chimpanzee (*Pan Troglodytes*)," *Behaviour*, vol. 60 (1977), pp. 179–94.

44. Hanby and Brown, "The development of sociosexual behaviors."

45. J. Colvin, "Influences of the social situation on male emigration," in *Primate Social Relationships*, ed. R. A. Hinde (Sunderland, Mass.: Sinauer, 1983), pp. 160–71.

46. Elizabeth A. Missakian, "Genealogical mating activity in free-ranging groups of rhesus monkeys (*Macaca mulatta*) on Cayo Santiago," *Behaviour*, vol. 45 (1973), pp. 224–40.

47. Nash, "Development of the mother-infant relationship."

48. Packer, "Inter-troop transfer."

49. Ibid.

50. Savage and Malik, "Play and socio-sexual behaviour."

51. Pusey, Schumacher Stankey, and Goodall, "Age changes in frequency of mating."

52. C. B. Clark, "A preliminary report on weaning among chimpanzees of Gombe National Park, Tanzania," in *Primate Bio-social Development*, ed. S. Chevalier-Skolnikoff and F. E. Poirier (New York: Garland, 1977), pp. 235–60.

53. Pusey, Schumacher Stankey, and Goodall, "Age changes in frequency of mating."

54. Ibid.

55. Savage and Malik, "Play and socio-sexual behaviour."

56. Ibid.

57. Pusey, Schumacher Stankey, and Goodall, "Age changes in frequency of mating."

58. Goodall, "The behaviour of free-living chimpanzees."

59. Pusey, Schumacher Stankey, and Goodall, "Age changes in frequency of mating."

60. Jeffrey R. Walters, "Transition to adulthood," in *Primate Societies*, ed. B. B. Smuts, D. L. Cheney, R. M. Seyfarth, T. T. Struhsaker, and R. W. Wrangham (Chicago: University of Chicago Press, 1987), pp. 358–69.

61. Alberts, "Paternal kin discrimination."

62. Hans Kummer, *Social Organization of the Hamadryas Baboons* (Chicago: University of Chicago Press, 1968).

63. H. Sigg, A. Stobla, J. J. Abegglen, and V. Dasser, "Life history of hamadryas baboons," *Primates*, vol. 23 (1982), pp. 473–87; J. J. Abegglen, *On Socialization in Hamadryas Baboons* (Cranbury, New Jersey, 1984).

4 *Explaining the Westermarck Effect*

OR, WHAT DID NATURAL SELECTION SELECT FOR?

Arthur P. Wolf

> You are thinking of your sons—but do not you know that
> of all things upon earth, *that* is the least likely to happen;
> brought up, as they would be, always together like brothers
> and sisters? It is morally impossible. I never knew an
> instance of it. It is, in fact, the only sure way of providing
> against the connection.
> —Jane Austen, *Mansfield Park*

This is the argument with which Mrs. Norris persuaded her brother-in-law, Sir Thomas, to invite his niece, Fanny Price, to live at Mansfield Park. Mrs. Norris was being meddlesome, as usual, and her argument was self-serving. But was she right? Is *"that"* the thing least likely to happen when male and female children are brought up together? Is early association a good way, if not necessarily the only good way, of ensuring against a later "connection"?

Until the mid-twentieth century, custom in most of China (as well, I might add, as in most of Korea) gave families a choice of how to acquire wives for their sons. One way was to wait until the son was fully grown and then arrange a marriage with a young adult who would come to live with her husband and his parents. In this case, which I call "major marriage," the young couple did not ordinarily meet until the day of their wedding. The alternative was what I call "minor marriage."[1] In this case the family "adopted" (or bought) a girl and raised her as a daughter-in-law. Many of these girls were taken in as infants and nursed by their future mother-in-law. It was common practice for a woman who bore a son and then a daughter to give her daughter away and raise her son's wife in her daughter's place.[2] In Taiwan these girls were called *sim-pua*, "little daughters-in-law."

Much of the argument of this paper and some of the data appear in a paper entitled "Reformulating (Yet Again) the Westermarck Hypothesis, or Was Dr. Ellis Right?"

Although the great majority of girls adopted as *sim-pua* were taken by couples who already had a son for the girl to marry, there were exceptions. Prompted by the belief that adopting a girl would enhance the wife's chances of bearing a son—or by anxiety about finding a girl to adopt in a tight adoption market—some couples adopted a little daughter-in-law before producing a son.[3] The result was that about 5 percent of the women in minor marriages were present to witness their husband's birth. We will see later that they are the key to answering an important part of the question raised by Mrs. Norris's meddling.

When I began field research in northern Taiwan in the late 1950s, I found that nearly half of the women over thirty years of age had been married in the minor fashion. It was what I saw of their lives that alerted me to the significance of this form of marriage, but the evidence I present in this chapter is largely drawn from household registers compiled in Taiwan by the Japanese colonial government. In 1905 the Japanese police interviewed every person on the island and recorded, among other things, their birth date, their adoption date if adopted, and the form and date of their current marriage. After that, household and village heads were required to report to the police all vital events—births, deaths, marriages, divorces—within ten days of their occurrence.[4] My previous work shows that the great majority of this information is highly reliable. The only exceptions are events in the lives of people who were already elderly when the police interviewed them in 1905. In this chapter, I will avoid the problems this creates by confining my analysis to the marriages of women born after 1890.

My work to date—which now spans more than forty years—has established three points relevant to the question of whether *that* is unlikely when children are brought up "always together." The first two are based on the reconstruction of upwards of 20,000 marriages. The first point is that when women in minor marriages were adopted at an early age, their fertility was 40 percent *lower* than that of women in major marriages.[5] The second point is that when women in minor marriages were adopted early, their chances of experiencing divorce were three times *higher* than those of women in major marriages.[6] My third point is based on interviews concerning extramarital sexual relations among 551 women. My finding is that according to their relatives and neighbors, women who married a childhood associate were more than twice as likely to seek sexual satisfaction outside of marriage than women who married a stranger.[7]

When I began publishing this evidence in the early 1960s, the great majority of social scientists were convinced that Mrs. Norris was wrong. With very few exceptions, they accepted Sigmund Freud's contention that "an *incestuous love-choice* is in fact the first and regular one, and that it is only later that any opposition is manifested towards it, which is not to be sought

in the psychology of the individual."[8] This attitude changed remarkably in
the following thirty years. Studies of couples reared together in Israel,
Lebanon, and Taiwan convinced all but the most conservative Freudians
(and unreconstructed social constructionists) that Mrs. Norris was right.[9]
The man who had listened to the Mrs. Norrises of the world reemerged as
the authority on the subject of incest avoidance. It is now generally accepted
that Edward Westermarck was right when, in 1895, he argued that "there is
a remarkable absence of erotic feelings between people living closely together from childhood."[10]

The problem now is why? Why does early association inhibit sexual attraction? The evidence reviewed in this volume by Alan H. Bittles (Chapter
2) and Anne Pusey (Chapter 3) argues that from an evolutionary perspective, the most likely answer is that the dangers of close inbreeding selected
for something that causes us to respond to early association with an enduring sexual aversion. But what is this something? What did natural selection
select for? Mrs. Norris insists that even if Fanny Price turns out "to have
the beauty of an angel," she "will never be more to [her cousins] than a sister."[11] But why? Why would growing up together cause two young men to
ignore a beautiful young woman?

In this case, as in most cases of the kind, there is, as Karl Popper put it,
"the problem of formulating the problem."[12] I think the most promising
formulation is to ask what it is about "being always together" that matters.
In other words, what conditions are necessary to produce "the Westermarck effect"? Thus my strategy will be to try to answer three questions
that Westermarck neglected. The first is, what does *childhood* mean in his
famous hypothesis? How early do males and females have to meet to qualify as "living closely together from childhood"? The second is, what does
closely mean in the phrase "living closely together"? What kind of association is necessary to produce sexual aversion? And the third is, are males
and females equally sensitive to the effects of early association? In a review
of the fifth edition of Westermarck's *The History of Human Marriage*, the
famous sexologist Havelock Ellis suggested that the inhibition is "probably
more clearly marked in the female" than in the male.[13] Westermarck's only
response was to note that "Dr. Ellis may be right."[14] But was he?

In this chapter I will pursue these questions by examining more closely
than I have previously variation in sexual attraction *among* minor marriages. My measures of sexual attraction are general marital fertility and an
especially constructed index I call the fertility/divorce index. General marital fertility is calculated by dividing the number of births among married
women aged fifteen to forty-five by the number of years they were married
during these ages. My fertility/divorce index is general marital fertility ad-
justed to take account of the number of divorces experienced by the women

in question. The adjustment consists of subtracting from the numerator of the marital fertility rate five births for every divorce. The number five is arbitrary. I have chosen it as a rough estimate of the number of children women who were divorced would have borne if they had not been divorced.

The fertility/divorce index is intended to capture in one number the two clearest quantifiable manifestations of the sexual aversion aroused by early association. Divorce alone is an unreliable measure because it was stigmatized—to the point of being regarded as polluting—and therefore infrequent. We have to assume that the great majority of the couples affected by early association never considered divorce as the solution to an unhappy relationship. My reason for creating an index that takes it into account is the possibility that the few couples who did risk divorce were those who were the most sensitive to the inhibiting effects of early association. The fertility/divorce index is best considered an attempt to estimate what the fertility rate would have been if there were no divorce.

The marriages out of which my tables and graphs were constructed include all the minor marriages contracted by women born in the years 1890 to 1920 in eighty-five villages and neighborhoods (including the neighborhoods that constitute two small towns). Twenty-eight of these communities were included in the analyses presented in my 1995 book and are located in northwestern Taiwan (in the southwestern corner of the Taipei Basin and in the hills overlooking Hsin-chu City). The remainder are additions to my database and are located in the Pescadores Islands and in northeastern Taiwan (on the coast south of Keelung City and on the Ilan Plain).

I will begin with what *childhood* means—or should mean—in the Westermarck hypothesis. This is not simply a question of when childhood ends. It is also a question about the relative impact of the ages included. One possibility is that the Westermarck effect is primarily a product of association during a particular developmental phase. The other is that it is largely a function of the number of years of association prior to puberty. To address the matter, I have plotted my two indexes of sexual attraction against the age at which women in minor marriages first met their future husband. The plotted points include all the data available except for marriages in which the wife was adopted before the husband's birth and marriages in which the husband was eight or more years older than the wife. We will see later why these two were excluded.

Table 4.1 and Figure 4.1 display in numerical and visual form the relationship between the fertility of minor marriages and the wife's age at adoption. It is important to note that the data are heavily concentrated at age zero and diminish steadily as age rises. This is because Taiwanese women preferred to adopt their son's wife as early as possible. The result for us is that above age four the plotted relationship is somewhat irregular. But if we

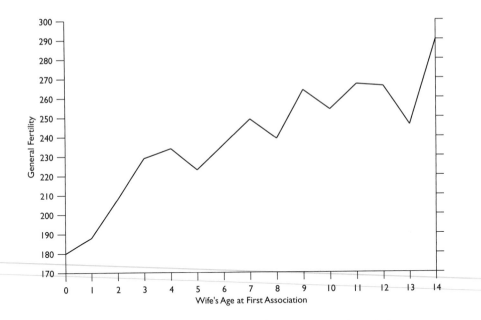

FIGURE 4.1. General Fertility by Wife's Age at First Association

TABLE 4.1
General Fertility by Wife's Age at First Association

Wife's Age at First Association	Years of Marriage	General Fertility Rate
0	28,873	180
1	5,514	188
2	3,698	208
3	2,651	229
4	1,927	234
5	2,560	223
6	1,886	236
7	1,473	249
8	1,867	239
9	1,834	264
10	1,930	254
11	1,765	267
12	1,743	266
13	1,335	246 } 268
14	1,042	290 }

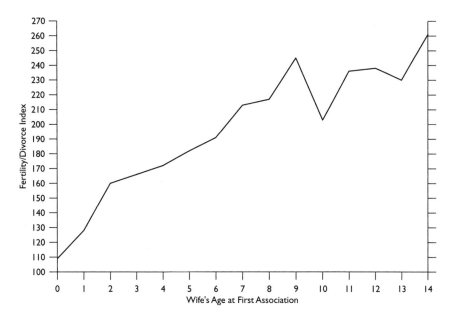

FIGURE 4.2. Fertility/Divorce Index by Wife's Age at First Association

TABLE 4.2
Fertility/Divorce Index by Wife's Age at First Association

Wife's Age at First Association	Years of Marriage	Fertility/Divorce Index
0	28,873	109
1	5,514	128
2	3,698	160
3	2,651	166
4	1,927	172
5	2,560	182
6	1,886	191
7	1,473	213
8	1,867	217
9	1,834	245
10	1,930	203
11	1,765	236
12	1,743	238
13	1,335	230 ⎫ 245
14	1,042	261 ⎭

average the figures for ages thirteen and fourteen (which is, I think, justified by the small denominators), the trend is sharply upward for every two-year period before age nine, which marks the beginning of a high, fairly level plateau. We must note, however, that the rise up to age three is far steeper than the rise from age three to age nine. The average rise per year is 16.3 units for ages zero to two, as compared with 5.8 units for ages three to nine.

The data displayed in Table 4.2 and Figure 4.2 surprised me. I did not expect adjusting for divorce to make such a large difference. Not only is the distance between the lowest and highest points in Figure 4.2 54.6 percent greater than in Figure 4.1, the relationship between the plotted variables is more regular. There are no reversals before age nine. Nonetheless, our two indexes of sexual dissatisfaction present broadly similar profiles. They rise sharply in early childhood, modulate, and then level out well before adolescence. The only significant difference is that the fertility/divorce index locates the end of the early childhood rise a year earlier than the general fertility index.

Although more data are needed to stabilize the trends at later ages, that shown is sufficient to provide a better-than-tentative answer to our first question. What does *childhood* mean in the Westermarck hypothesis? How early do males and females have to meet to qualify as "living closely together from childhood"? The crude answer is "before age ten." The more refined answer is that while every year of association before age ten adds to the sum of the Westermarck effect, association beginning before age three is particularly potent. There is—to use Patrick Bateson's language—a sensitive period and a very sensitive period.[15]

The data presented above say that the earlier a girl destined to marry in the minor fashion was adopted, the less satisfied she was with the relationship created by her marriage. But what about her husband? Wasn't his age when he first met his future wife a factor? Girls adopted as infants were most often matched with a boy born two or three years previously, but many were matched with boys born six or seven years previously and others with boys born after their arrival. Might it be, then, that we need to attend to the husband's age at first association as well as the wife's? It all depends on whether Dr. Ellis was right. If the "instinct" was really "more marked in the female" than in the male, it might not matter very much how old a boy was when his future wife was adopted.

One of my reasons for expanding my database beyond that employed in my 1995 book was to be able to address this question. A large database is required because to examine the effect of the husband's age at first association, one has to control on the wife's age. Even with the incremental data I have to confine my analysis to marriages in which the wife was adopted as an infant (i.e., before age one). The data I use include all minor marriages in which the wife was adopted as an infant except those cases in which she was adopted before her husband was born. Again I leave these special cases until later.

The relationships between my two indexes and the husband's age at his wife's adoption are shown in Tables 4.3 and 4.4 and Figures 4.3 and 4.4. Again the profiles presented are broadly similar. Both indexes decline from age zero to age one, remain low until age seven, rise sharply at age eight, and then decline sharply at age twelve. I will ignore for the time being both the early and late declines because the figures are not well grounded. Thus what the evidence says is that so long as the husband is not eight or more years older than his wife, his age when he first meets her does not matter. The rise in both indexes if he is eight or more years older is most likely due to the couple's not "living closely together from childhood." A man eight or more years older than his wife was nine or more when she joined his family. He would already be spending most of his waking hours studying or working rather than playing with the nursing infant destined to be his wife.

Must we conclude, then, that Dr. Ellis was indeed right? If we assume, as many anthropologists and most biologists do assume, that females invest far more in their offspring than males, parental investment theory suggests that he should be right.[16] But is this assumption correct? It is certainly the case that when the wife is the younger partner in a minor marriage, the trajectory of my indexes is largely controlled by her age at first association. The husband's age does not matter as long as he is still a small child. But what if the wife were the older partner? It is this question that makes important those marriages in which the wife was adopted before the husband's birth. They allow us to see what happened when the husband was a nursing infant at first association and the wife a small child.

The relevant data are displayed in Tables 4.5 and 4.6. The figures shown there are irregular because their base is small, but they all support the same conclusion. They say that when the husband is the infant at first association the aversion is as strong as when the wife is the infant. The critical comparison is between the data in Tables 4.3 and 4.4 (where the wife is an infant and the husband is a small child) and that in Tables 4.5 and 4.6 (where the husband is an infant and the wife is a small child). To make the comparison easier, I have created Table 4.7, in which the two sets of data are shown side by side. The obvious fact is that early association on the husband's part is as consequential as early association on the wife's part. It is particularly worth noting that when an infant boy is matched with a three- to six-year-old girl, fertility is even lower than when an infant girl is matched with a three- to six-year-old boy.

Suppose for a moment that the figures in Tables 4.5 and 4.6 rose steadily with age, as is the case with the figures in Tables 4.1 and 4.2. One could

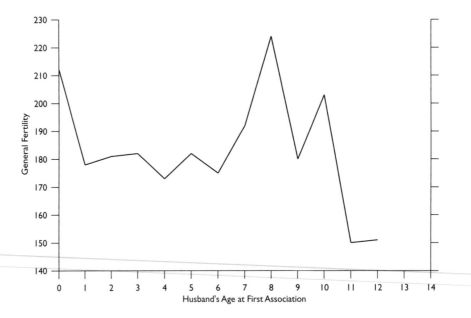

FIGURE 4.3. General Fertility by Husband's Age at First Association

TABLE 4.3
General Fertility by Husband's Age at First Association
When Wife's Age at First Association Is Zero

Husband's Age at First Association	Years of Marriage	General Fertility Rate
0	554	212
1	1,911	178
2	6,146	181
3	6,725	182
4	5,397	173
5	3,629	182
6	2,339	175
7	2,084	192
8	1,390	224 ⎤
9	451	180 ⎟
10	558	203 ⎬ 182
11	307	150 ⎟
12	317	151 ⎦
13	—	—
14	—	—

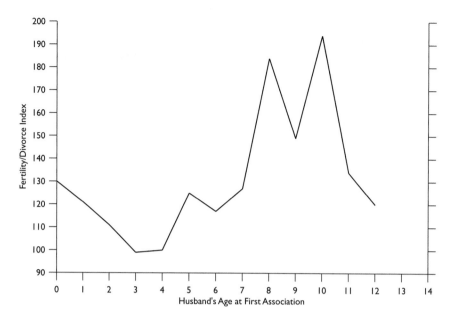

FIGURE 4.4. Fertility/Divorce Index by Husband's Age at First Association

TABLE 4.4
Fertility/Divorce Index by Husband's Age at First
Association When Wife's Age at First Association Is Zero

Husband's Age at First Association	Years of Marriage	Fertility/Divorce Index	
0	554	130	
1	1,911	121	
2	6,146	111	
3	6,725	99	
4	5,397	100	
5	3,629	125	
6	2,339	117	
7	2,084	127	
8	1,390	184	
9	451	149	
10	558	194	156
11	307	134	
12	317	120	
13	—	—	
14	—	—	

then argue that these figures are low only because females are sensitive to early association with a younger male as well as early association with an older male. In other words, one could argue that Dr. Ellis was right. But in fact there is no significant trend among the figures in either table. The reason can only be because the source of the problem in these marriages is the male. His age does not vary, and so the indexes do not vary. Consequently, I conclude that Dr. Ellis was wrong. Males and females are equally sensitive to the sexually inhibiting effects of early association. What really matters is whether the male or the female is the younger partner. An interesting—and, I think, important—implication is that we have either to reject parental investment theory or to conclude that our male ancestors made a much larger contribution to their offspring's survival than is commonly assumed.

I skipped the second of the questions raised above in order to pursue the implications of Tables 4.3 and 4.4 for the question of a sex difference. The question skipped was, What does *closely* mean in the phrase "living closely together"? All I have to contribute to the answer at this point is the data reported in Tables 4.3 and 4.4. The indexes in both tables indicate that a minor marriage is less likely to manifest sexual aversion if the husband is eight or more years older than the wife. This suggests that *closely* means intense interaction of the kind commonly found among children who are close in age and who regularly play together. Examination of the composition of the sibling sets in which the parties to minor marriages were reared might prove the point. The problem is that it would require a database several times the size of the one on which this chapter rests.

Contrary to Mrs. Norris's confident prediction, Fanny Price fell in love with the younger of her two cousins, Edmund. He ignored her for years in favor of Mary Crawford, but in the end he ceased "to care about Miss Crawford, and became as anxious to marry Fanny, as Fanny herself could desire."[17] So once again Mrs. Norris's advice turned out to be bad advice, but we can now see that though misapplied, the general principle on which she based herself was sound. When she moved to Mansfield Park, Fanny Price was "just ten years old."[18] She had just reached the age at which association with the opposite sex is no longer inhibiting. Mrs. Norris's "*that*" was to be expected.

We need, then, to revise the Westermarck hypothesis. Age at first association—and particularly the age of the younger member of any pair of potential partners—is far more important than previously realized. My revision—phrased to remind us of the author of the original hypothesis—reads, "There is a remarkable absence of erotic feelings between people who live together and play together before age ten. The absence is particularly marked among couples brought together before age three, and, for any given couple, largely depends on the age of the younger partner when they first meet."

TABLE 4.5

*General Fertility by Wife's Age at First Association
When Wife Is Adopted Before Husband's Birth*

Wife's Age at First Association	Years of Marriage	General Fertility Rate
0	213	188
1	822	202
2	1,949	200
3	1,508	167
4	1,151	177
5	470	147
6	340	168

TABLE 4.6

*Fertility/Divorce Index by Wife's Age at First Association
When Wife Is Adopted Before Husband's Birth*

Wife's Age at First Association	Years of Marriage	Fertility/Divorce Index
0	213	127
1	822	128
2	1,949	131
3	1,508	83
4	1,151	109
5	470	85
6	340	50

TABLE 4.7

*Comparison of the Consequences of Early Association
When the Wife Is an Infant (Tables 4.3 and 4.4)
and the Husband Is an Infant (Tables 4.5 and 4.6)*

Age of Spouse at First Association	General Fertility		Fertility/Divorce Index	
	Wife Is an Infant	Husband Is an Infant	Wife Is an Infant	Husband Is an Infant
0	212	188	130	127
1	178	202	121	128
2	181	200	111	131
3	182	167	99	83
4	173	177	100	109
5	182	147	125	85
6	175	168	117	50

How, then, are we to explain the reformulated Westermarck effect? Why does association before age ten inhibit sexual attraction? Why is association during the first two or three years of life particularly inhibiting? Why is it that the strength of the inhibition depends on the age of the younger partner rather than on the age of both partners? In sum, what is it that natural selection selected for that produces these effects?

Answering these questions will not be easy because it requires facing up to a dilemma that I have so far avoided mentioning. It stems from the fact that when the husband in a minor marriage is the older partner, his age at first association does not influence the intensity of the aversion as long as he is still a child. The intensity depends entirely on the age of the wife at first association. This implies that the aversion is largely located on her side of the relationship. But how, then, could the fertility of minor marriages be so low and the divorce rate so high? My data come from China, where, in the words of Friar Domingo Navarette, "the wives are half slaves and their subjection is extraordinary."[19] How could wives who were "half slaves" frustrate their husbands' sexual drive and their desire for a houseful of sons and grandsons? How could women whose subjection was "extraordinary" get a divorce when divorce was regarded as a social disaster? I think that anyone who knows the traditional Chinese family will agree that they couldn't. The aversion must have been felt, acutely, by the husband as well as by the wife. But how could this be, if the strength of the aversion depends on the wife's age at first association and not on the husband's age? This is the dilemma. Solving it is one of the feats the Hercules of incest avoidance must perform. Sadly, the field presently lacks a credible Hercules, but there are three promising candidates.

In a recent book one of my Stanford colleagues, Eleanor Maccoby, argues that "there is a powerful tendency for children to segregate themselves by gender in childhood and to play more compatibly with same-sex partners." This "drift into same-sex groups" is found in all societies and "begins to show itself in the third year of life . . . and progressively strengthens until it is strong indeed by middle childhood."[20] "It may be, then," Maccoby suggests, noting the reaction to minor marriages, "that children's spontaneous avoidance of cross-sex others who are not kin serves the biological function of keeping these others within the pool of potential mates."[21]

The timing Maccoby describes fits neatly with the age-dependent trends displayed in Tables 4.1 and 4.2 (and, more dramatically, in Figures 4.1 and 4.2). It also fits with the fact that when the wife is the younger partner, the strength of the aversion in minor marriages depends on the age at which she joins her future husband's family. Being the younger of the two, she would be the last to join an exclusive same-sex group. Consequently, her age, but not her husband's age, would influence how frequently the couple interacted as children. The argument does not explain the Westermarck effect, but it does explain why the effect is age-related and why, when the female is younger than the male, her age seems to matter more than his.

The argument suggested by Maccoby is entirely compatible with Westermarck's emphasis on "living closely together from childhood." A more radical approach to the problem of incest avoidance has been suggested to me by another of my Stanford colleagues, Hill Gates, who, as an anthropologist with a special interest in Taiwan, knows minor marriages at first hand.[22] Gates's argument is radical in suggesting that what produces the Westermarck effect may not be some aspect of playing together, eating together, and sleeping together. It may be something far more easily identified. It may be having been breastfed by the same woman.

Gates's argument begins with the well-known finding that both males and females prefer as sexual partners persons whose major histocompatibility complex (MHC) is different from their own.[23] The reason, it is argued, is that they smell like relatives. Might it not be, then, that the dangers of inbreeding have selected for the ability to identify relatives by their odor and avoid them? The problem with this elegant solution to the problem of incest avoidance is that it is contradicted by all the evidence presented above. Couples married in the minor fashion were not relatives and therefore should not have been olfactorily obstructed. In a recent article Mark Schneider and Lewellyn Hendrix have tried to salvage the MHC hypothesis by suggesting that we learn the odors of the people with whom we are reared and avoid them because they are probably relatives.[24] Gates avoids the complications this introduces by arguing that because the development of MHC is strongly influenced by breastfeeding, children who are breastfed by the same women tend to smell alike even if they are not siblings.

The great advantage of Gates's hypothesis is that it resolves the minor marriage dilemma. Taiwanese men were always put off by girls their mother nursed for the simple reason that they had been nursed by the same woman. Another advantage of the hypothesis is that it explains why association during the first two years of life was so much more potent than association during later years. This is because girls adopted at an early age were often nursed by their future mother-in-law, while those adopted later were not. The one problem with the hypothesis is that it does not explain why association beginning after age three has any impact at all.

The third candidate is the one I nominated in my 1995 book, *Sexual Attraction and Childhood Association*.[25] It begins with John Bowlby's account of what he calls "attachment behavior."[26] Bowlby defines this "as any form of behavior that results in a person's attaining proximity to some other differentiated and preferred individual, usually conceived of as stronger and/or wiser." The behavior includes clinging, crying, calling, greeting, and smiling.

It is evident from six months onward, "when an infant shows by his behavior that he discriminates sharply between his mother-figure, a few other familiar people, and everyone else." It reaches its "maximum during the second and third year of life and then diminishes slowly."[27]

Bowlby believes attachment behavior is one of three basic components of human nature. The second is "the urge to explore the environment, to play and to take part in varied activities with peers," and the third is caregiving, which Bowlby characterizes as "the prime role of parents and complementary to attachment."[28] Just as human beings are born with a tendency to seek and maintain contact with persons who are better able to cope, so also they are born with an innate tendency to succor and support other human beings who are not yet able to cope. This is, in Bowlby's view, "readily understood since it serves to promote the survival of offspring and thus the individual's own genes."[29]

My solution to the dilemma posed by minor marriages is to combine Bowlby's argument with Westermarck's. The selection forces favoring the dispositions underlying attachment and caregiving push us ever closer to our genetic relatives and thus expose us ever more acutely to the dangers of inbreeding, while the selection forces favoring the dispositions underlying incest avoidance push us ever further from our genetic relatives and thus make us ever more vulnerable to the dangers of isolation. Consequently, if the advantage to be gained by strengthening one set of dispositions is to result in a net genetic gain, the other set must be strengthened at the same time. To evolve at all, the two sets of dispositions had to evolve together. The fact that attachments and sexual aversions both form more readily before age three than after is not coincidental. They are the same thing.[30]

What is proposed, then, is that "little daughters-in-law" taken before age three attached themselves to their future husband because he was older and appeared "stronger and/or wiser." This behavior elicited caregiving in return and thereby created an asexual relationship because having evolved together with incest avoidance, attachment and caregiving are inherently contrasexual. The reason the fertility of minor marriages varies with the wife's age and not the husband's is simply because in most cases the wife is the younger partner and thus the one who does or does not form an attachment. The location of the aversion moves from the female side to the male side when the husband is the younger partner.

These hypotheses need to be tested and can be tested. The Taiwan household registers contain all the information needed to reconstruct the exact composition of households and even neighborhoods. Thus it would be possible to determine the availability of same-sex playmates and the likelihood that couples in minor marriages had grown up separated by membership in same-sex groups. The registers also record the removal of infants by death and adoption—even when the infant is only a few days old. Thus one could determine whether a woman was lactating when she adopted a little daughter-in-law and thereby estimate the chances of her having nursed her son's wife. The problem is that analyzing a sufficiently large number of household registers would be a huge task. Even Hercules would have hesitated.

NOTES

1. The relative frequency of these two forms of marriage in Taiwan is documented in Arthur P. Wolf and Chuang Ying-chang, "Marriage in Taiwan, 1881–1905: An example of regional diversity," *Journal of Asian Studies*, vol. 54 (1995), pp. 781–95.

2. A detailed account of major and minor marriages as institutions is available in Arthur P. Wolf and Chieh-shan Huang, *Marriage and Adoption in China, 1845–1945* (Stanford, Calif.: Stanford University Press, 1980).

3. See Wolf and Huang, *Marriage and Adoption*, chapter 18, pp. 242–50.

4. The registers are described in detail in Wolf and Huang, *Marriage and Adoption*, chapter 3, pp. 16–33.

5. See Arthur P. Wolf, *Sexual Attraction and Childhood Association: A Chinese Brief for Edward Westermarck* (Stanford, Calif.: Stanford University Press, 1995), chapters 7 and 12, pp. 115–34 and 198–213.

6. Ibid., chapters 6 and 11, pp. 98–114 and 181–97.

7. Ibid., chapters 5 and 10, pp. 79–97 and 166–80.

8. Sigmund Freud, *A General Introduction to Psychoanalysis* (1920), trans. Joan Riviere (New York: Pocket Books, 1953), pp. 220–21.

9. See, most importantly, Yonina Talmon, "Mate selection in collective settlements," *American Sociological Review*, vol. 29 (1964), pp. 491–508; Joseph Shepher, "Mate selection among second-generation kibbutz adolescents: Incest avoidance and negative imprinting," *Archives of Sexual Behavior*, vol. 1 (1971), pp. 293–307; Justine McCabe, "FBD marriage: Further support for the Westermarck hypothesis," *American Anthropologist*, vol. 85 (1983), pp. 50–69; and Wolf, *Sexual Attraction*.

10. Edward Westermarck, *A Short History of Human Marriage* (London: Macmillan and Co., 1926), p. 80.

11. Jane Austen, *Mansfield Park* (1814) (New York: Penguin Books, 2000), p. 8.

12. Karl Popper, *Unended Quest: An Intellectual Autobiography* (La Salle, Illinois: Open Court Publishing, 1976), pp. 134–35.

13. Havelock Ellis, *Views and Reviews* (London: Desmond Harmsworth, 1932), p. 168. First published in the *Birth Control Review* (New York), September 1928.

14. Edward Westermarck, "Recent theories of exogamy," *Sociological Review*, vol. 26, no. 1 (1934), p. 36.

15. Patrick Bateson and Robert A. Hinde, "Developmental changes in sensitiv-

ity to experience," in *Sensitive Periods in Development*, ed. Marc H. Bornstein (Hillsdale, N.J.: Lawrence Erlbaum, 1987), pp. 19–20.

16. See Robert L. Trivers, "Parental investment and sexual selection," in *Sexual Selection and the Descent of Man*, ed. Bernard Campbell (London: Aldine, 1972), pp. 136–79.

17. Austen, *Mansfield Park*, p. 387.

18. Ibid., p. 12.

19. *The Travels and Controversies of Friar Domingo Navarette*, ed. J. S. Cummins, Hakluyt Society, ed. ser., no. 118 (Cambridge: Cambridge University Press, 1960), vol. 1, p. 161.

20. Eleanor E. Maccoby, *The Two Sexes* (Cambridge: Harvard University Press, 1998), pp. 29–30.

21. Ibid., p. 94.

22. Personal communication during dinnertime conversation. Hill Gates is my wife.

23. See C. Wedekind, T. Seebeck, F. Beetens, and A. J. Paeike, "MHC-dependent mate preferences in humans," *Proceedings of the Royal Society of London*, Series B, vol. 260 (1995), pp. 345–49; and C. Wedekind and S. Furi, "Body odour preferences in men and women: Do they aim for specific MHC combinations or simply heterozygosity?" *Proceedings of the Royal Society of London*, Series B, vol. 264 (1997), pp. 1471–79.

24. Mark A. Schneider and Lewellyn Hendrix, "Olfactory sexual inhibition and the Westermarck effect," *Human Nature*, vol. 11, no. 1 (2000), pp. 65–91.

25. See Wolf, *Sexual Attraction*, chapter 28, pp. 463–75. My argument was anticipated by Mark Erickson in articles published several years previously. See Mark T. Erickson, "Incest avoidance and familial bonding," *Journal of Anthropological Research*, vol. 45., no. 3 (1989), pp. 270–80; and "Rethinking Oedipus: An evolutionary perspective on incest avoidance," *American Journal of Psychiatry*, vol. 150, no. 1 (1993), pp. 400–413.

26. Essentially the same explanation was developed independently by Mark T. Erickson. See Erickson, "Incest avoidance and familial bonding."

27. John Bowlby, "Attachment theory, separation anxiety, and mourning," in *American Handbook of Psychiatry*, ed. David A. Hamburg and H. K. H. Brodie (New York: Basic Books, 1975), vol. 6, p. 292.

28. John Bowlby, *A Secure Base* (New York: Basic Books, 1988), pp. 163–65.

29. Ibid., p. 165.

30. This argument is developed in detail in Wolf, *Sexual Attraction*, chapter 28, pp. 439–75.

5 *Ancient Egyptian Sibling Marriage and the Westermarck Effect*

Walter Scheidel

Under Roman rule, all residents of Egypt were required to participate in a periodic census held every fourteen years. The household was the
basic unit of registration; each head of a household had to file a return list-
ing all of its members with their names, ages, and kinship affiliations. A
minute fraction of the millions of census declarations ever filed has survived on papyrus. So far, close to 300 of them have been published and
subjected to demographic analysis.[1] Of 121 current marriages documented
in these records, twenty are between full siblings and four between half-
siblings. Another three sibling unions are known to have ended in divorce.
Other kinds of papyrus texts outside this corpus contain references to thirteen further sibling couples, one of them divorced. All these unions were
(legally) monogamous.

Sibling marriage, attested for the second and the early third centuries AD,
appears to have been more prevalent in urban settings than in the villages.
In the most amply documented location, the district capital of Arsinoe in the
Fayum Oasis (southwest of modern Cairo), seventeen of forty-six known
unions, or 37 percent, are between full siblings. Owing to the limited availability of suitable sibling-spouses in any particular family and a strong preference for younger wives, the observed incidence approaches the feasible
maximum. In this environment, sibling marriage was a cultural norm rather
than merely an acceptable option.[2] This invalidates Arthur Wolf's claim,
based on an inadequate sample of evidence, that according to the Egyptian
data, "even in the absence of an incest taboo, children who are raised together rarely marry."[3]

Brother-sister marriage in Roman Egypt has long been noted as a con-
spicuous exception to near-universal taboos against regular sexual and marital relations between very closely related kin. Traditionally, this phenomenon has been discussed from a cultural-constructivist perspective.[4] In a
series of interlocking studies, I have sought to redress the balance by analyz-

ing the available data within a biosocial framework. Rather than guessing at possible explanations, I have focused on the probable repercussions of and constraints on this peculiar custom, drawing on comparative evidence for inbreeding depression and avoidance behavior.[5] As a result, Roman Egyptian sibling marriage has now finally entered the debate on the biological basis of human incest taboos.[6]

It is easy to show that the most intensely incestuous segments of the Egyptian population, such as the residents of Arsinoe, were more heavily inbred than any other known human population.[7] In the absence of direct or indirect textual or physical evidence, the scale of resultant inbreeding depression is uncertain but likely to have been substantial (see below). Explicit references to an aversion against sibling marriage are also missing from the record (such as private letters on papyrus or literary accounts). In this regard, the Egyptian case differs from Zoroastrian "close-kin marriage" between parents and children and brothers and sisters, practiced in Iran and the Near East in antiquity and the Middle Ages, perhaps primarily in polygamous circles. Glorifying incest as exceptionally meritorious, the Zoroastrian spiritual authorities expressly considered it challenging even for devoted insiders and envisioned only infrequent sexual encounters of this nature. Instinctive reluctance was reported by sympathetic insiders and hostile outsiders alike.[8]

Comparative evidence sheds some light on the probable psychological effects of Egyptian sibling marriage. Research on modern populations has established a "remarkable absence of erotic feelings between people living closely together from childhood"[9] and consequent aversion against reproductive sexual intercourse at mature ages, known as the "Westermarck effect." In his work on the custom of "minor marriage" in Taiwan—in which the parents of a newborn son adopted (or bought) a young girl, often an infant, to be raised together with their son as his future wife—Wolf found that the marital fertility of girls who had been adopted at an early age was 40 percent lower than that of wives who had been raised by their own families, and also that "minor marriages" with a small age gap between spouses were more than twice as likely to involve adultery and three times as likely to end in divorce than other unions. Wolf shows that the intensity of early childhood association was the critical variable mediating marital success.[10] In this chapter, I will assess the significance of this factor in the context of Egyptian sibling marriage.

Data

Table 5.1 summarizes the vital statistics of all Roman Egyptian sibling couples that are currently known. The majority of cases (twenty-seven

TABLE 5.1
Sibling Couples Attested in Papyrus Documents from Roman Egypt
(first to third centuries AD)

	Couples			Offspring Residing with Parents	
Case	Age Gap (years)	Age/Husband (years)	Age/Wife (years)	Number	Ages (years)
Full siblings (census returns)					
1	2	31	29	1	8
2	3	21	18	1	?
3	3	46	43	6 or 7	20, 16, 14, 10, 7, 3, 2(?)
4	4	[4]4	40	2	12, 8
5	4	46	42	5	12, 10, 8, 6, 2
6	<6	5[-]	54	8	29, 26, 23, ?, 17, ?, 9, 7
7	6	36	30	0	
8	8	21	13	0	
9	8	21	13	0	
10	8	22	14	0	
11	8	44	52	1	8
12	? (0–9)	20	2[-]	0	
13	10	48	38	1	?
14	17	34	17	1	1
15	?	21	?	2	>1, <1
16	?	29	?	2	1, 1 (twins)
17	?	?	35	2	?
18	?	40	?	1	2
19	?	73	?	2	40, ?
20	?	?	?	3	23, ?, ?
21*	?	?	?	2	?, 10
22	?	?	?	2	?, ?
Full siblings (other documents)					
23	1	?	?	1 known	?
24	3	?	?	2 known	?
25*	4	54	50	?	?
26	8	43	35	1 known	22
27	?	?	?	2 known	?
28	?	?	?	1 known	~14
29	?	?	?	1 known	13
30	?	?	?	1 known	4
31	?	?	?	1 suggested	?
Half-siblings (census returns)					
32	2	31	29	4	14, 6, 3, 2(?)
33*	4	22	26	3	?
34	12	30	18	0	
35	22(?)	[6?]3	41	4	21, 13, 11, ?
36	?	?	?	2	?, 14
Full or half-siblings (other documents)					
37	?	?	?	1 known	
Half-siblings (other documents)					
38	?	?	?	1 known	8
39	?	?	?	2 known	
40*	?	?	?	?	

KEY: * divorced; ? missing information; (?) uncertain reading; [ ] missing/illegible numeral
SOURCES: Cases 1–22 and 32–36: Roger S. Bagnall and Bruce W. Frier, *The Demography of Roman Egypt* (Cambridge: Cambridge University Press, 1994); cases 23–24, 27: P. Amherst 75; case 25: P. Kron. 52; case 26: *BGU* 183 (*P.* 6867); cases 28, 37: *P. Tebt.* 320; case 29: *P. Oxy.* 43.3096; case 30: *WChr.* 211; case 31: *P. Duk.* inv. 491; case 38: *P. Pet.* 1–2; case 39: *P. Oxy.* 43.3137; case 40: *P. Strasb.* 768.

of forty) are attested in census returns. Thanks to the standardized format of the census declarations and their objective to record all coresident kin in a given household, the demographic information contained in these documents is often more complete and more fully contextualized than that preserved in the other texts. Even so, each census return gives only a snapshot of household composition at a particular moment in time. Thus, no completed life histories are available; while young couples would have produced additional offspring later on, some of the children of older couples may already have left the parental home at the time of registration and remain invisible. Deceased progeny is not recorded.

In view of very high levels of mortality—most likely of the order of 30 percent up to age one, 45 percent up to age five, and 55 percent up to age twenty (and probably even more in those cases in which inbreeding depression needs to be taken into account; see below)—the discrepancy between the number of children ever born to a couple and the number of their children who were alive at any given time must have been substantial. As a consequence, the census records cannot be used to measure the total marital fertility rate (TMFR). Other texts that mention brother-sister marriage are frequently concerned with one particular child (listed as "known" in Table 5.1) and omit others that may have existed, as well as information on parental age. In a number of instances, the papyrus texts are so poorly preserved that some entries are fragmentary or difficult to decipher; these problems are indicated in Table 5.1.

In the census returns in general, illegitimate children are usually identified as such, yet there are no references to the (acknowledged) illegitimacy or adoption of sibling-spouses or their descendants. Although unknown or concealed adulterous or extramarital conceptions may sometimes have occurred, there is nothing to suggest that sibling-spouses were frequently biologically unrelated to one another and/or their putative children. No practice comparable to Chinese "minor marriage" is known from Roman Egypt. In fact, individual full sibling-spouses are repeatedly listed as "brother/sister from the same father and the same mother."[11] With a single exception (case 37), this precision enables us to distinguish between couples of full and half-siblings.

By the standards of other ancient age records or census counts in modern developing countries, the degree of age rounding (defined as preference for numerals ending in multiples of five) in the census returns is remarkably low, and particularly modest in those urban locales that produced much of the relevant texts.[12] The age data in Table 5.1 show no significant deviation in favor of multiples of five (thirteen of eighty-five final digits, or 15.3 percent, consistent with the expected rate of 20 percent). For this reason, the recorded age gaps between siblings may generally be accepted at face value.

Birth Intervals and Spousal Age Difference

The observation that early childhood association curbs sexual attraction and may even trigger sexual aversion at mature ages provides an obvious starting point for a reappraisal of brother-sister marriage. I therefore agree with Lewellyn Hendrix and Mark Schneider that "a crucial consideration would be how intimate siblings were in Graeco-Roman Egypt during the sensitization period of early childhood."[13] However, Seymour Parker is mistaken in claiming that "there are no data on the closeness of opposite-sex sibling socialization during the early years."[14] The recorded age gaps between sibling spouses shed some light on the likely level of variation in the intensity of early childhood association in this group. Two different ways of calculating the average age difference between full sibling spouses produce very similar results.

The average length of birth intervals can be deduced from the probable fertility schedule of married women in Roman Egypt. The general age distribution in the census returns as well as comparative evidence from early-twentieth-century Egypt and comparable historical populations points to a mean life expectancy at birth of between twenty and twenty-five years.[15] Again drawing on the census returns, Bruce Frier was able to show that the attested birthing schedule is fully consistent with a natural fertility regime, characterized by the absence of family limitation achieved through stopping behavior; thus, birth intervals were relatively long but childbearing continued into the forties, gradually decreasing with diminishing female fecundity.[16] Positing a mean life expectancy at birth of about twenty-two to twenty-five years, this observation allowed Frier to reconstruct two versions of the Roman Egyptian marital fertility schedule, one reflecting the raw data and the other based on an idealized Gompertz curve fitted to the documented maternal age distribution of childbearing (Table 5.2).

According to the model pattern, a married woman at the peak of her fecundity on average gave birth every 3 years. This rate slowed to one birth every 3.5 years in her thirties before dropping sharply in her forties. Approximately seven out of eight children born in wedlock were born to women aged fifteen to thirty-nine and were on average separated by 3.4 years. Two qualifications deserve attention. First, divorce and the death of husbands would sometimes interrupt periods of high fertility, thereby increasing the average actual distance between births. The total fertility rate (TFR) was close to six children per woman surviving to menopause, fully 30 percent lower than TMFR in the model. Second, and even more important, high levels of prereproductive mortality would, on average, greatly raise the age difference between siblings surviving to sexual maturity. If only half of all newborns could hope to survive into their early teens, as

TABLE 5.2
Marital Fertility in Roman Egypt
(mean number of births per maternal age cohort)

Age	Attested	Gompertz Model
12–14	0.066	0.069
15–19	1.16	1.245
20–24	1.715	1.665
25–29	1.835	1.625
30–34	1.465	1.495
35–39	1.09	1.31
40–44	1.095	0.83
45–49	0.67	0.185
Total	9.096	8.424

SOURCE: Adapted from Bruce W. Frier, "Natural Fertility and Family Limitation in Roman Marriage," *Classical Philology* 89 (1994), 325, table 1.

seems likely, the mean age gaps between these individuals would have been twice as wide as the average birth intervals. For children born to women aged fifteen to thirty-nine, the resultant age difference would, on average, have amounted to approximately seven years.

This schematic calculation tallies well with the attested age gaps between sibling spouses in Table 5.1. The mean age difference for sixteen couples of full siblings is 6.06 years, while the median (including case 6) is 6 years. If four half-sibling couples are included, the mean rises to 6.85 years. These observed rates are fully consistent with the estimate generated by the model fertility schedule.

Five papyrus documents report brother-sister marriage in two successive generations, and in one case, the practice continued across three generations.[17] In those instances, the probable impact of inbreeding depression also needs to be taken into account. Frier's fertility schedule is predicated on the assumption of zero or marginal natural population growth. However, owing to the likely impact of inbreeding depression (expressed in higher rates of fetal loss, child mortality, and disability), it remains unclear whether incestuous families were able to reproduce at full replacement level. If they did, they may on average have needed more live births to produce the same number of mature offspring than nonkin couples. In an earlier study, I tentatively calculated the required increase in marital fertility, which amounts to 19 percent for first-generation sibling-spouses, 30 percent in the second generation, and 42 percent for the last of three successive sibling matings.[18] For first-generation sibling couples (arguably the most common variety), this boosts TMFR from 8.4 to 10 and reduces average birth intervals accordingly, to about 2.9 years in the fifteen- to thirty-nine-year maternal age bracket. At the same time, owing to higher prereproductive wastage, the mean age difference between mature siblings remains unchanged. Alternatively, it is very well possible that, on average, incestuous couples could not reproduce at full replacement level. In that case, mean birth intervals may have been closer to the norm, while the resultant average age gaps between surviving siblings would have exceeded seven years.

The overall incidence of second-generation sibling marriage remains obscure. On average, no more than about 60 percent of all brothers had a sister of marriageable age, and owing to a strong preference for younger wives (in sixteen of eighteen sufficiently well-documented couples in Table 5.1, the brother-husband is older than the sister-wife), the actual incidence of sibling marriage in any given generation must have been closer to 30 percent. Thus, any increase of mean spousal age difference through inbreeding depression was unlikely to affect more than a relatively small minority of all sibling couples. We may conclude that the average age gap between full sibling spouses must have been of the order of seven years.

This overall average necessarily conceals considerable variation. In our data set, nine of twenty-one sibling couples, or 43 percent, are separated by eight or more years of age. Despite an emerging consensus on the relevance of early childhood cohabitation and cosocialization for mating preferences in adulthood, there is some disagreement as to the precise length of the sensitization period. Joseph Shepher reckoned with mature avoidance caused by close contact during the first six years of life.[19] The most detailed studies, Wolf's work on "minor marriages," show a strong sensitization effect during the first three years followed by a steady decrease. When a future husband was eight years or older at the time when a newborn girl was adopted as his future wife, the impact of their belated association on subsequent marital success (measured by fertility and divorce rates) was fairly negligible.[20] This configuration is equivalent to the relationship between a brother-husband and a sister-wife who was eight or more years younger than he. In their analysis of modern incest, Irene Bevc and Irwin Silverman find that full genital—that is, potentially reproductive—intercourse between siblings today is strongly associated with prolonged separation during the first three years of life.[21]

All these studies agree that the effects of early childhood association on later mating behavior are strongest for the first few years of life and marginal if sibling age difference exceeds six to eight years. Hence, in almost half of all cases, the Egyptian sibling spouses were too far apart in age to have been exposed to intensive sensitization. The Chinese and other comparative data indicate that under these circumstances (which include strong cultural expectations favoring incest), instinctive aversion to reproductive relations would have been weak or altogether missing and that the success of such sibling unions would have been unimpaired by the Westermarck effect. Moreover, the consistent preference for younger sister-wives must have limited close contact between older brothers and significantly younger sisters; while older girls may have been expected to care for infant brothers, the reverse scenario is considerably less likely.

Conjugal Dissolution

The Chinese data for "minor marriages" suggest that divorce rates were a function of the degree of early childhood association between future spouses.[22] Thus, the likelihood of conjugal dissolution was highest when the wife had been adopted between ages zero and four, and negligible after age nine, while for men, it was highest when the future spouse had been adopted during the first six years of the boy's life.[23] This raises the question of whether a similar trend can be discerned in Roman Egypt. Strictly speaking, the age at adoption in "minor marriages" is not fully equivalent to age difference between biological sibling-spouses; in the Chinese context, both future partners would sometimes initially spend some time on their own before the girl was introduced into her new family. In one Chinese sample, 44.5 percent of these girls were adopted before age one and 64.3 percent before age three.[24] Thus, while age at adoption is generally indicative of the intensity of early childhood association, it can serve only as a rough proxy for age difference between siblings as observed in the Egyptian data.

Nevertheless, even allowing for this discrepancy, the Chinese evidence would seem to predict elevated rates of marital failure among Egyptian sibling couples close in age, especially among those who were separated by no more than four to six years. On the face of it, the data are consistent with this prediction (Table 5.3).

The average divorce rate for nonsibling couples in the census returns is 11.8 percent (13 of 110).[25] By contrast, 30 percent of known unions between siblings who were close in age ended in divorce. Owing to the exiguous size of the sample, the significance of this deviation from the putative mean remains very weak ($p < 0.17$, z-test). However, two qualifications are in order. The final divorce rate remains unknown even for these few couples. The lack of completed life histories means that the available data systematically underestimate the actual incidence of separation. This raises the possibility that complete life histories would reveal a more significant correlation between age difference and divorce. Besides, the suggestive match between the patterns of divorce in the scant Egyptian and the much more numerous Chinese data would make it seem rash to dismiss the findings derived from the smaller sample out of hand.

TABLE 5.3
Cases of Divorce by Spousal Age Gap

	Age Gap Between Spouses	
	1–4 Years	6+ Years
All couples	10	10
Divorced couples	3	0

SOURCE: Table 5.1.

Cross-Fostering

The data discussed so far are consistent with comparative evidence and general predictions derived from it; thus, Egyptian sibling marriage may have worked reasonably well when spousal age difference was considerable but less so if spouses were close in age. However, several prolific sibling couples with a modest age difference diverge from this inherently plausible pattern (cases 3, 5, 6, 32). The Chinese data as recalculated by Wolf show a strong correlation between age at adoption (our rough proxy for age difference) and marital fertility even if divorce is controlled for; close association in the first few years of life results in significantly reduced reproductive success at mature ages.[26] The four Egyptian instances confound our expectations that the same ought to have been true of genuine sibling marriage. Several explanations are possible. The paucity of data may be an obvious answer; in a sample this small, some random anomalies would be unsurprising. Perhaps more important, the Chinese evidence leaves no doubt that although the impact of early childhood association on fertility was significant, it was also limited in scale. In many cases, inhibitions were to some extent overcome by parental coercion and social expectations.[27] As a consequence, above-average fertility in some sibling couples need not be particularly noteworthy.

For all that, the observed age gaps (and implied birth intervals) in these successful families may provide a clue to a more satisfying answer that takes full account of the mechanisms of early childhood association (Table 5.4). The three most prolific couples in this group, all of them consisting of full siblings, list living offspring with a total of thirteen known age gaps between siblings, ranging from one to four years. (The two unknown intervals in case 6 are likely to fall in the same range.) Some of these intervals seem short for two reasons. First, we have to allow for infant and early childhood mortality; some of the actual birth intervals may have been shorter than the observed age gaps between living siblings. (The first interval in case 32 is an obvious candidate.) Second, birth intervals of one or two years are hard to reconcile with ancient Egyptian nursing practices. Judging by the literary

TABLE 5.4
*Birth Intervals of Offspring of Prolific Sibling Couples Separated
by Fewer Than Six Years of Age (in years)*

Birth Interval	Couples				
	Case 3	Case 5	Case 6	Case 32	All
1	1(?)	—	—	1(?)	2(?)
2	1	3	1	—	5
3	1	—	2	1	4
4	3	1	—	—	4
<6	—	—	1	—	1
<8	—	—	1	—	1
8	—	—	—	1	1

SOURCE: Table 5.1.

tradition, children were to be breast-fed for three years.[28] The papyrological record from the Roman period includes forty wet-nursing contracts in which unrelated women (often slaves) are hired to breast-feed an infant (often a slave or foundling, sometimes the child of the employer) for a specified period of time. The term of employment is known in twenty-eight cases; 2 and 2.5 years are the most common periods, while 3.5 years is the longest. Terms for freeborn nurslings range from 2 to 3 years.[29]

"It is now well-established that breastfeeding can exert a powerful contraceptive effect, potentially delaying the return of full fecundity during the postpartum period by two years or more."[30] The evidence cited here shows that freeborn Egyptian children were regularly supposed to be breast-fed for up to three years. In addition to this strong cultural preference for prolonged breast-feeding, other factors contributed to the comparatively long birth intervals suggested by Table 5.2. Thus, endemic ill health, parasitism, and high mortality at all ages, even among the propertied classes, must have resulted in correspondingly elevated rates of fetal loss and temporary sterility,[31] as did inbreeding depression in those couples who were themselves the issue of brother-sister unions (see above).

There is no direct evidence that any known sibling-spouse had been cared for by a wet nurse. Even so, the observed combination of extended breast-feeding, short age gaps between living siblings (with potentially even shorter birth intervals), and the apparent popularity of wet-nursing arrangements suggests that some of the children of the most prolific sibling couples may have been nursed by women other than their own mothers. At the very least, intervals of one or two years as documented in Table 5.4 strongly support this assumption. This in turn raises the possibility that such couples may have followed the example of their own parents and that some sibling-spouses had themselves been in the care of wet nurses. After all, age gaps of one or two years are likewise attested for some sibling couples (cases 1, 23, and 32).

The available evidence for the technical details of wet-nursing arrangements is somewhat ambiguous. In a number of cases, the contracts obligate the wet nurse to raise the nursling in her own home (or that of her owner if she was a slave).[32] This clause, however, is found only in agreements providing for the nursing of slaves and foundlings and does not appear in any known contract for freeborn infants. Instead, in one text of the latter category, the wet nurse expressly promises her employer, "I will stay day and night in your house together with the baby." The "day and night" stipulation is also found in another contract for a free child. A poorly preserved, lacunose contract for a third free nursling appears to give the wet nurse an option of whether to follow her employer beyond his regular residence, which implies that she would ordinarily stay there.[33]

An infant's transfer to the home of a wet nurse would have sheltered the child from early association with siblings and other family members. However, the contracts suggest that when their own children were involved, parents preferred to hire a live-in wet nurse. While it is impossible to be sure that future sibling-spouses were never kept in the homes of unrelated wet nurses, there is no positive evidence for this. In any case, the "day and night" stipulation makes it likely that a live-in wet nurse acted as a child minder beyond the actual feeding process and would have been a major focus of attachment for the child. The same would have been true of lactating household slaves who took care of their owners' children, a practice that was common in Rome and must also have existed in affluent Egyptian circles.[34]

Recent work on olfactorily mediated sensitization in early childhood raises the possibility that wet-nursing arrangements of this kind might have interfered with the development of inhibitions to sexual intercourse with close relatives at mature ages. It is now well known that various animal species including humans are highly sensitive to information, conveyed by body odor, on genetic variation in the major histocompatibility complex (MHC, known as HLA in humans). The MHC, a highly polymorphic group of genes, serves as a matching system used by the immune system to discriminate between self and others and is therefore also instrumental in kin recognition.[35] Olfactory perception of minute genetic differences in MHC-type can be shown to influence mating preferences in animals and humans.[36] This mechanism frequently—though not invariably—guides individuals toward mates with MHC types that are different from their own. In such cases, it is heterozygosity as such that is favored rather than any particular MHC type.[37] This generalized preference for heterozygosity may enhance immunological resistance to pathogens.[38] At the same time, it also favors outbreeding and may therefore have evolved to avoid inbreeding.[39]

Humans can be shown to prefer the body odor of potential mates with different MHC types.[40] Despite these preferences, evidence for disassortative mating (in the form of marriage) has so far remained ambiguous.[41] It may be possible to explain the absence of disassortative mating practices with reference to a lack of free mate choice or the very real possibility that MHC preference is based on (less precise) familial imprinting rather than self-reference.[42]

Preference for different MHC strains is the result of an early learning process. At the very least, associate reference (in which infants are sensitized to their own MHC type by being sensitized to the body odor of those around them, who are most likely to be close kin) plays a significant role alongside self reference (in which MHC preference would be directly determined by one's own body odor). The importance of associate reference was demonstrated when laboratory mice that had been removed from their natal litter after birth, fostered by an unrelated female, and then returned to their own group exhibited avoidance of the MHC type of their foster parents but not of their biological kin.[43] The same effect could be shown in wild-derived mice in seminatural conditions.[44]

I know of no empirical evidence for the impact of cross-fostering on the sensitization of humans. In view of other structural similarities with regard to MHC-mediated preferences across different mammalian species including humans, it seems at least possible that human infants might react similarly to the way mice react. In pseudo-sibling relationships, such as Chinese "minor marriage," exposure to the body odor of biologically unrelated members of the adoptive family can be expected to trigger subsequent avoidance of mates from within that group and so help account for the Westermarck effect.[45] On occasion, Chinese mothers were known to breast-feed their adopted daughters-in-law; a girl nursed by the mother of a biologically unrelated brother-spouse may thus have been sensitized against later sexual relations with that male.[46]

In Roman Egypt, the opposite effect may have occurred. Several years of regular exposure to the breast milk and the breast and axillary odor of an unrelated wet nurse may have sensitized small children to an MHC type other than their own and thereby reduced their inhibitions against sexual relations with their own kin at mature ages. (In this scenario, it would not matter whether two sibling-spouses had been nursed by the same stranger or by different women.) The overall impact of this sensitization relative to concurrent sensitization to coresident siblings remains open to debate. If future sibling-spouses had been physically removed from their natal families for the first few years of their lives, the dominance of their sensitization to nonkin MHC-types could hardly be doubted. As it is, it remains uncertain whether breast-feeding and nursing were more potent elements of early childhood sensitization than contact with coresident siblings and parents. Further studies on the effects of cross-fostering are required to shed light on this issue.

Conclusion

We cannot say why so many couples in Roman Egypt arranged marriages between their own children. Evidence of religious motivation, which is richly available for Zoroastrian close-kin marriage, is lacking here.[47] Concern about the preservation of family property and family privilege has been mooted as a possible motive but could equally well have been alleviated by the unexceptional custom of first-cousin marriage. (Unions of this type also appear in the census returns but can be identified only if a couple's sibling parents resided in the same household.) Both status enhancement via religious devotion and status preservation via "closed-family marriages" would constitute plausible proximate mechanisms to raise fitness even if they were bound to be maladaptive in the long run.

As I have tried to show, it may be easier to explain the actual workings of this custom in Darwinian terms. There is no sign that a short-lived memetic mutation such as monogamous sibling marriage rendered evolved constraints immaterial. Rather, its temporary success appears to have been both facilitated and limited by a combination of different factors. In a substantial proportion of all cases, probably close to one-half, considerable age differences between the spouses would have reduced or removed inhibitions to sexual intercourse and reproduction. Some couples who were close in age may conceivably have been sensitized to the body odor of unrelated wet nurses and might consequently have been spared strong feelings of sexual aversion at mature ages. In still other cases, sibling-spouses who were close in age and had been sensitized to their own kin may well have experienced elevated rates of conjugal dissolution. All in all, there is nothing to show that as far as the correlation between early childhood association and sexual inhibition is concerned, the evidence for Roman Egyptian sibling marriage deviates significantly from the pattern derived from the Chinese data on "minor marriages" and other information on the demographic context of incestuous behavior and incest avoidance in humans.

NOTES

1. Roger S. Bagnall and Bruce W. Frier, *The Demography of Roman Egypt* (Cambridge: Cambridge University Press, 1994).

2. Walter Scheidel, "Brother-Sister Marriage in Roman Egypt," *Journal of Biosocial Science*, vol. 29 (1997).

3. Arthur Wolf, *Sexual Attraction and Childhood Association: A Chinese Brief for Edward Westermarck* (Stanford, Calif.: Stanford University Press, 1995), p. 433.

4. For example, Keith Hopkins, "Brother-Sister Marriage in Roman Egypt," *Comparative Studies in Society and History*, vol. 22 (1980), pp. 303–54; Brent D. Shaw, "Explaining Incest: Brother-Sister Marriage in Greco-Roman Egypt," *Man*, vol. 27 (1992), pp. 267–99; S. Parker, "Full Brother-Sister Marriage in Roman Egypt: Another Look," *Cultural Anthropology*, vol. 11 (1996), pp. 362–76.

5. Walter Scheidel, *Measuring Sex, Age, and Death in the Roman Empire: Explorations in Ancient Demography* (Ann Arbor, Mich.: Journal of Roman Archaeology, 1994), pp. 9–51; Scheidel, "Brother-Sister and Parent-Child Marriage Outside Royal Families in Ancient Egypt and Iran: A Challenge to the Sociobiological View of Incest Avoidance?" *Ethology and Sociobiology*, vol. 17 (1996), pp. 319–40; Scheidel, "Brother-Sister Marriage in Roman Egypt."

6. L. Hendrix and M. A. Schneider, "Assumptions on Sex and Society in the Biosocial Theory of Incest," *Cross-Cultural Research*, vol. 33 (1999), pp. 193–218; M. A. Schneider and L. Hendrix, "Olfactory Sexual Inhibition and the Westermarck Effect," *Human Nature*, vol. 11 (2000), pp. 65–91.

7. Scheidel, "Brother-Sister Marriage in Roman Egypt."

8. Scheidel, "Brother-Sister and Parent-Child Marriage in Ancient Egypt and Iran"; "Brother-Sister and Parent-Child Marriage in Premodern Societies," in *Human Mate Choice and Prehistoric Marital Networks*, ed. Kenichi Aoki and Takeru Akazawa (Kyoto: International Research Center for Japanese Studies, 2002), pp. 33–47.

9. Edward Westermarck, *A Short History of Human Marriage* (London: Macmillan, 1926), p. 80.

10. Wolf, *Sexual Attraction*; and Chapter 4 of this volume.

11. The most recently published text (case 31) documents the first known instance of marriage between twins: N. Gonis, "Incestuous Twins in the City of Arsinoe," *Zeitschrift für Papyrologie und Epigraphik*, vol. 133 (2000), pp. 197–98.

12. Scheidel, *Measuring Sex, Age, and Death*, pp. 53–91.

13. Hendrix and Schneider, "Assumptions on Sex and Society," p. 213.

14. Parker, "Full Brother-Sister Marriage," pp. 373–74.

15. Bagnall and Frier, *Demography of Roman Egypt*, 75–110; Walter Scheidel, *Death on the Nile: Disease and the Demography of Roman Egypt* (Leiden, Netherlands: Brill, 2001), pp. 118–80.

16. Bruce W. Frier, "Natural Fertility and Family Limitation in Roman Marriage," *Classical Philology*, vol. 89 (1994), pp. 318–33.

17. Scheidel, *Measuring Sex, Age, and Death*, pp. 11–12.

18. Ibid., 25, table 1.11, based on K. Ralls, J. D. Ballou, and A. Templeton, "Estimates of Lethal Gene Equivalents and the Cost of Inbreeding in Mammals," *Conservation Biology*, vol. 2 (1988), pp. 185–93; Alan H. Bittles and James V. Neel, "The Costs of Human Inbreeding and Their Implications for Variations at the DNA Level," *Nature Genetics*, vol. 8 (1994), pp. 117–21.

19. Joseph Shepher, *Incest: A Biosocial View* (New York: Academic Press, 1983), pp. 51–62.

20. Wolf, *Sexual Attraction*; and Chapter 4 of this volume.

21. I. Bevc and I. Silverman, "Early Separation and Sibling Incest: A Test of the Revised Westermarck Theory," *Evolution and Human Behavior*, vol. 21 (2000), pp. 151–61; cf. already their "Early Proximity and Intimacy Between Siblings and Incestuous Behavior: A Test of the Westermarck Hypothesis," *Ethology and Sociobiology*, vol. 14 (1993), pp. 171–81.

22. Wolf, *Sexual Attraction*; chap. 11.

23. Ibid., p. 197; and Chapter 4 of this volume.

24. Wolf, *Sexual Attraction*, p. 170.

25. Bagnall and Frier, *Demography of Roman Egypt*, p. 123.

26. Wolf, Chapter 4 of this volume.

27. Hendrix and Schneider, "Assumptions on Sex and Society," p. 202.

28. Joyce Tyldesley, *Daughters of Isis: Women of Ancient Egypt* (London: Penguin Books, 1995), p. 69.

29. Madriadele Manca Masciadri and Orsolina Montevecchi, *I contratti di baliatico* (Milan: private printing, 1984), pp. 32–35.

30. James W. Woods, *Dynamics of Human Reproduction: Biology, Biometry, Demography* (New York: Aldine de Gruyter, 1994), p. 370.

31. Scheidel, *Death on the Nile*.

32. Masciadri and Montevecchi, *I contratti*, p. 24.

33. Ibid., pp. 43, 149, 146.

34. Keith R. Bradley, "Wet-Nursing at Rome: A Study in Social Relations," in *The Family in Ancient Rome: New Perspectives*, ed. Beryl Rawson (Ithaca, N.Y.: Cornell University Press, 1984), pp. 201–29.

35. J. L. Brown and A. Eklund, "Kin Recognition and the Major Histocompatibility Complex: An Integrative Review," *American Naturalist*, vol. 143 (1994), pp. 435–61.

36. D. Penn and W. Potts, "How Do Major Histocompatibility Genes Influence Odor and Mating Preferences?" *Advances in Immunology*, vol. 69 (1998), pp. 411–36.

37. C. Wedekind and S. Füri, "Body Odour Preferences in Men and Women: Do They Aim for Specific MHC Combinations or Simply Heterozygosity?" *Proceedings of the Royal Society of London, Series B—Biological Sciences*, vol. 264 (1997), pp. 1471–79.

38. V. Apanius, D. Penn, L. R. Ruff, and W. K. Potts, "The Nature of Selection on the Major Histocompatibility Complex," *Critical Reviews in Immunology*, vol. 17 (1997), pp. 179–224.

39. D. Penn and W. Potts, "The Evolution of Mating Preferences and Major Histocompatibility Complex Genes," *American Naturalist*, vol. 153 (1999), pp. 145–64.

40. C. Wedekind, T. Seebeck, F. Bettens, and A. Paepke, "MHC-Dependent Mate Preferences in Humans," *Proceedings of the Royal Society of London, Series B—Biological Sciences*, vol. 260 (1995), pp. 245–49; Wedekind and Füri, "Body Odour Preferences."

41. For positive results, see C. Ober, L. R. Weitkamp, N. Cox, H. Dytch, D. Kostyu, and S. Elias, "HLA and Mate Choice in Humans," *American Journal of Human Genetics*, vol. 61 (1997), pp. 497–504 (and cf. E. Genin, C. Ober, L. Weitkamp, and G. Thomson, "A Robust Test for Assortative Mating," *European Journal of Human Genetics*, vol. 8 [2000], pp. 119–24). For negative findings, see P. W. Hedrick and F. L. Black, "HLA and Mate Selection: No Evidence in South Amerindians," *American Journal of Human Genetics*, vol. 61 (1997), pp. 505–11; Y. Ihara, K. Aoki, K. Tokunaga, K. Takahashi, and T. Juji, "HLA and Human Mate Choice: Tests on Japanese Couples," *Anthropological Science*, vol. 108 (2000), pp. 199–214; cf. also K. Jin, T. P. Speed, and G. Thomson, "Tests for Random Mating for a Highly Polymorphic Locus: Application of HLA Data," *Biometrics*, vol. 51 (1995), pp. 1064–76.

42. Cf. Penn and Potts, "Major Histocompatibility Complex Genes," p. 416.

43. K. Yamazaki, G. K. Beauchamp, K. Kupniewski, J. Bard, L. Thomas, and E. A. Boyse, "Familial Imprinting Determines H-2 Selective Mating Preferences," *Science*, vol. 240 (1988), pp. 1331–32; A. Eklund, "The Effect of Early Experience on MHC-Based Mate Preferences in Two B10.W Strains of Mice (*Mus domesticus*)," *Behavior Genetics*, vol. 27 (1997), pp. 223–29. Cf. also K. F. Arcaro and A. Eklund, "A Review of MHC-Based Mate Preferences and Fostering Experiments in Two Congenic Strains of Mice," *Genetica*, vol. 104 (1998), pp. 241–44.

44. D. Penn and W. Potts, "MHC-Disassortative Mating Preferences Revised by Cross-Fostering," *Proceedings of the Royal Society of London, Series B—Biological Sciences*, vol. 265 (1998), pp. 1299–1306.

45. Schneider and Hendrix, "Olfactory Sexual Inhibition."

46. Wolf, *Sexual Attraction*, p. 183; H. Gates in Chapter 8 of this volume.

47. Scheidel, "Brother-Sister and Parent-Child Marriage in Ancient Egypt and Iran"; "Brother-Sister and Parent-Child Marriage in Premodern Societies."

6 *From Genes to Incest Taboos*

THE CRUCIAL STEP

Neven Sesardic

Not in Our Genes?

Today the idea that an evolutionary approach may be fruitful for research in the social sciences is being passionately defended by some and no less passionately contested by others. The resistance to Darwinism comes mainly in two distinct varieties. The first type of criticism is based on empirical or methodological objections against the current attempts to use evolutionary considerations to throw some light on social science explananda. The other line of opposition, however, is much harder to pin down and discuss because it is fueled more by rhetoric than by argument. It defines itself, rather vaguely, as a fight against "biological reductionism" and "genetic determinism" and is often accompanied by slight (or not so slight) ideological overtones. In this chapter, I will deal only with the former (methodological) kind of criticism. But since I don't want to leave the latter, hazily antireductionist source of opposition to biology without comments, and since I don't know how to approach it in a serious way, let me wiggle out by presenting to you a rhymed parody, "Gene-mania," that captures some of the more ideological criticism's characteristic flavor:

GENE-MANIA

Who today is not sick and tired
Of all those guys so gene-inspired?
They find a gene for every this or that:
For being gay, smart, alcoholic, fat . . .
We have to stop that madness. Take no offense,
But this approach doesn't make much sense.
True, some fools thought after Watson-Crick
That genes could really do the trick.
What they sought they did not find—
Those ill-fated biologists of the mind.

Loonies still insisting on double helix
Must be cured of that *idée fixe*.
Now we know better, we've been instructed:
Human phenotypes are socially constructed.
"Darwinism, yes," we exclaim with glee,
Adding a proviso: "Only to some degree!"
It is quite all right for a fly or bird,
There the gene-talk is not at all absurd.
But, wait, don't rush to generalize
From these creatures of smaller size.
Looking at *Drosophila melanogaster*,
Please, don't read too much into it, buster.
Well, they rub their genitals, no doubt,
But that's nothing to get hot about;
You must be completely off your tracks
If you think that what they have is—sex.
Your anthropomorphism and sex obsession
Deserve of course our full compassion;
Yet for the prejudices so amazing
You badly need some consciousness-raising.
Besides, to tell you frankly, there's another thing,
It all too much smacks of—hmm, the right wing.
Hence in Boston, Stanford or Minneapolis
We might well need the thought police.
Why? Because all this genetic chitter-chatter
Is certainly not a laughing matter.
So be responsible, mind what you say.
Danger! You are talking DNA!

Incest as a Test Case

To what extent is evolutionary biology relevant for social science? The battle over this question is being fought on many fronts, and the discussion encompasses a number of specific topics. The issue of incest taboos stands out though. It is typically regarded as the critical case for evaluating theoretical aspirations of sociobiology and evolutionary psychology. For example, both in his book *Consilience* and in several subsequent interviews, Edward O. Wilson illustrates the success of sociobiology with the example of Westermarck's biological account of incest prohibitions.[1] Richard Dawkins also confirms the special place that this theme occupies in contemporary debates about biology and culture: "Although I usually resist the temptation to indulge in simple 'selfish gene' explanations of the social behavior of humans and other domestic animals, inbreeding avoidance is the one case for which I feel reasonably confident. 'The social science orthodoxy' has always seemed to me *particularly* daft in this area."[2] John Maynard Smith echoes Dawkins's thought: "To me, the most interesting question is how far evolutionary biology can contribute to the human sciences. As I have explained, I am a doubter. But I have been wrong on this issue before. Ten years ago I regarded incest avoidance as an entirely cultural phenomenon; only a bigot could hold this view today."[3]

Three Claims About the Aversion

In an earlier publication I suggested that Westermarck's theory about incest is best broken into three component claims:

1. A sexual aversion tends to develop between those raised together in early childhood.
2. This aversion is an evolutionary adaptation (serving as a barrier to inbreeding depression).
3. The aversion causes (expresses itself as) the incest prohibition.[4]

Bill Durham has usefully dubbed these three claims (1) the aversion hypothesis, (2) the adaptation hypothesis, and (3) the expression hypothesis.[5] Each of these claims says something about the sexual aversion: (1) The adaptation hypothesis says that the sexual aversion exists; (2) the adaptation hypothesis says that the cause of the sexual aversion is natural selection; and (3) the expression hypothesis says that the effect of the sexual aversion is the social prohibition of incest. I agree with Durham that the aversion hypothesis has the strongest empirical support and that the expression hypothesis is the weakest link in Westermarck's line of argument. One reason for problems with the expression hypothesis is that a lot of preliminary analysis and clarification has to be done before one can embark on the straightforward task of empirical evaluation or hypothesis testing. The complexity of these anthropological issues naturally attracts the visitors whose professed job is precisely to untangle difficult conceptual and methodological puzzles: philosophers. But scientists do not always warmly welcome philosophers to join the discussion. And, as it happens, this inhospitableness is not wholly unjustified.

Steven Weinberg once wrote, "The insights of philosophers have occasionally benefited physicists, but generally in a negative fashion—by protecting them from the preconceptions of other philosophers."[6] In the debate about incest as well, it seems to me that, unfortunately, some philosophical contributions have clouded the issues instead of elucidating them. I hope, however, that the analysis undertaken in the next section will be recognized as containing more than just this purely negative result (the criticism of other philosophers' views).

Inhibition and Prohibition: A Difference in Content?

The crucial idea in Westermarck's account of incest taboos is that it is the biological inhibition that gives rise to a cultural prohibition. It is quite clear that this hypothesis is fraught with difficulties and that many questions have to be answered before the argument that biological inhibition leads to cultural prohibition becomes persuasive. But some thinkers want to go much further than just pointing to that explanatory gap; they claim that a very general reason blocks in advance any transition from biological inhibition to cultural prohibition and that we should acknowledge that even without looking into empirical issues. Bernard Williams refers to this transcendental obstacle as the "representation problem."

Not only does extra conceptual content have to be introduced to characterize the human prohibition, but also the introduction of that content stands in conflict with the proposed biological explanation of it. . . . There are no sanctions against marrying those that one is brought up with (as such); the sanction is against marriages which would constitute close in-breeding. The conceptual content of the prohibition is thus different from the content that occurs in the description of the inhibition. It indeed relates to the suggested *function* of that inhibition, but that fact will not explain how the prohibition which is explicitly against in-breeding will have arisen. It certainly does not represent a mere "raising to consciousness" of the inhibition.[7]

Although Williams's idea reduces to a simple conceptual point, it still promises to settle an empirical issue; it is offered as a reason why we are entitled to dismiss a scenario in which cultural prohibition is seen as a mere manifestation of biological inhibition. Despite being quite influential and even making an impact on some scholars conducting empirical research on incest, this piece of a priori anthropology is fundamentally flawed. Since I have criticized it extensively elsewhere, let me here explain my reason for disagreeing with Williams more briefly and in a slightly different form.

Westermarck's basic idea is that for evolutionary reasons (i.e., the danger of inbreeding) a sexual aversion tends to develop between those who happened to be close associates in the crucial period of early childhood. Why does an aversion between childhood associates exist? Well, the hypothesis is that during the long time that natural selection has molded human psychology, close childhood associates were in fact almost always siblings. One obvious reason why it would be difficult for the aversion to be focused directly on siblings is that the concept of sibling is a fairly complicated social category—apparently an unlikely kind of object to which a genetically produced aversion might be immediately attached. Therefore, if there is a co-extensiveness (or very strong correlation) between these two properties, childhood associates (CA) and siblings (S), then natural selection might have found it easier, although no less effective as an instrument against in-

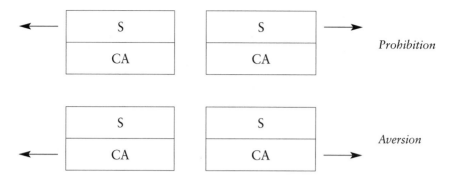

FIGURE 6.1. Aversion and Prohibition: Mismatch

breeding, to instill a sexual aversion between childhood associates (and not between siblings as such). Of course, the assumed strong correlation between CA and S would mean that as a consequence there would be a sexual aversion between siblings as well, but still, sensu stricto, it would only be correct to say that, in causal terms, what was producing the aversion was the characteristic CA, not S! This is readily confirmed in a pair of counterfactual or presumably rare exceptional situations in which the correlation between CA and S is violated: (1) As long as the CA-relation obtains, the aversion would still be there, even without S; (2) despite the presence of S, the aversion would not be there if the S in question failed to be CA.

On the basis of this (correct) description of the situation, Williams mounts an attack on the suggestion that the aversion could produce the incest prohibition. In Figure 6.1 the contrast between the aversion and the prohibition comes to the fore. Typically, two individuals of the opposite sex are *both* childhood associates (CA) *and* siblings (S), but the reasons they are kept apart are different in the two cases (aversion and prohibition).

Although as a rule it is the same individuals that are sexually kept away from one another by the aversion and by the prohibition, Williams wants us to notice that the forces of separation in these two cases are essentially different. The aversion is produced by a force opposing a CA-relationship, whereas what the prohibition condemns is just an S-relationship. Ergo, the argument goes, the prohibition cannot be a mere expression of the aversion because the two are directed to two entirely different aspects. The fact that CA-relationships are usually, or perhaps always, also S-relationships (and vice versa) is neither here nor there. Surely even after we concede that the two characteristics are correlated, it remains quite unclear by what kind of transformation the content of the prohibition that speaks only about S could be suddenly obtained from the mere aversion toward CA.

Williams's argument looks persuasive because it trades on a crucial ambiguity. The statement that the aversion is directed toward CA (childhood associates) can actually mean two things. First, in the *causal* sense, it can mean that the characteristic that is actually producing the aversion is CA. Second, in the *subjective* sense, it can mean that people who have the aversion experience it as being directed toward CA. I would like to point out that there is no necessary connection between the two senses; they are distinct and separable. In particular, if the aversion is directed toward CA in the causal sense, it by no means follows that the subjects would also experience it as being directed toward CA. They may well believe (wrongly) that the aversion is, say, generated by S.

The causal sense is relevant for the natural selection scenario. And yes, Westermarck's theory does contain a claim (the adaptation hypothesis) that the sexual aversion is objectively produced by being triggered by CA-aspect. The expression hypothesis, on the other hand, says that the prohibition is a manifestation of the inhibition. But here, in the context of the expression hypothesis, the content of the inhibition (that gives rise to the prohibition) is determined by the *subjective* sense of "being directed to," and *not* the *causal* sense! The fact that in the causal sense the aversion is indeed directed to CA (in contrast to the prohibition, which is directed to S) does not imply that there is a "mismatch" between the two contents, simply because the causal sense of the aversion's "being directed to" does not speak about content at all. It only speaks about which property is *objectively* causing the aversion. The content emerges only at a later stage when we ask how the people having the aversion experience it *subjectively*, that is, in terms of which property they conceptualize their own aversion. So it seems that Figure 6.1 was an oversimplification. It should be replaced by Figure 6.2.

Now we see that the fact that the aversion objectively picks out childhood associates as its object does not necessarily signify that its *content* is discrepant from the content of the prohibition (which forbids sex with siblings). For the content of the aversion is not fixed by such objective matters as the question about what is causing what. On the contrary, it essentially depends on how the aversion looks "from the inside" to those people who have it. And for all that matters, there is no guarantee here at all that there will be an accord between the objective and the subjective. In fact, given that natural selection found it easier to achieve its end (decreasing the probability of sex between siblings) by directing the aversion toward the more accessible and conveniently correlated proxy (childhood associates), it is perfectly possible that people, too, picking out environmental cues to make sense to themselves about their own aversion, also choose the line of least resistance and simply reach for the more meaningful, social category like *brother* or *sister*, rather than a seemingly irrelevant and queer charac-

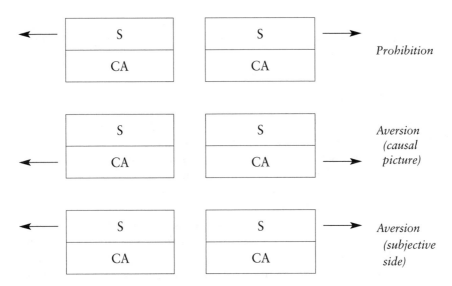

FIGURE 6.2. Aversion and Prohibition: No Mismatch

teristic like *someone with whom I spent the first years of my childhood.* Notice that I say that this is possible, not true. My concern here is not the truth, but the criticism of Williams's "impossibility proof." I tried to show that the transcendental obstacle he placed in the path of the Westermarckian account can be removed.

Williams hoped to demonstrate on very general grounds that the biological aversion couldn't produce the cultural prohibition. But his argument is based on the unjustified assumption that the content of the inhibition has to be presented to the subjects in terms of that property that is in fact causally operative in producing that inhibition. Once we realize that this is not necessary at all, the argument breaks down. The inhibition may well be subjectively presented differently (in a way that does not reflect its actual causal origin) and then, distorted in the "right" way, it may well match in content the corresponding cultural prohibition.

Westermarck's Argument as a Zero-Sum Game

There is another, again very general argument that is proposed as undermining the theory that the aversion gives rise to the prohibition (the expression hypothesis).

To make matters worse, there is little or no direct support for Westermarck's "moral disapproval" step. First, of the three studies already cited as supporting

Westermarck's argument about intimacy and aversion, not one shows evidence for a moral disapproval expressive of that emotion. . . . Any one of these cases might be taken as exceptional, and therefore dismissed from concern. But it is hard to accept that argument for all three, especially since the most familiar of all potential partners are not prohibited. In effect, the available evidence says, "Aversion, yes— moral disapproval, no."[8]

This is not a conceptual or transcendental argument. Durham's objection is based on probability. As he correctly notes, the existence of the Westermarckian sexual aversion between close childhood associates gets empirical support basically from three studies conducted in three different environments (Israel, Taiwan, and Lebanon). In all these cases there were clear signs of the lack (or decrease) of sexual attraction between childhood associates, although sexual contacts between the individuals in question were not prohibited but even encouraged. Durham then stresses one thing. These three studies are the best available evidence for the existence of the aversion. But in *none* of the three cases did the aversion lead to the prohibition. Why is that? He suggests that if the prohibition failed to emerge in only one of those situations, then it could still be reasonable to cling to the expression hypothesis and say that something exceptional occurred in that single instance that disrupted the usual tendency of the aversion to produce the prohibition. But if the same thing happened in *all* the three cases (as it did), he argues that this should make us strongly doubt, on purely inductive grounds, that there is any causal connection between the aversion and the prohibition.

I disagree. In my opinion, even the three repeated instances of "aversion, yes—moral disapproval, no!" do not justify skepticism with respect to the expression hypothesis. Contrary to Durham, I do not think that the three-fold presence of aversion-sans-disapproval points toward the conclusion that the former does not produce the latter. True, the fact that in those three quite vital cases for the evaluation of Westermarck's theory the aversion is *not* accompanied by moral disapproval cannot be dismissed as a mere coincidence (or "a few exceptions that prove the rule"). The absence of moral disapproval needs to be explained. One reason why the aversion is there, but moral disapproval is not, may indeed be that there is no intrinsic connection between the two and that in those other cases when they happen to be found together they are actually the result of different and largely independent causal processes. This is Durham's tack: if A is not followed by B in the three key cases, this undercuts (at least to some extent) the hypothesis that A produces B. But there is another way to understand why the aversion appears without moral disapproval in those three test cases. Maybe the absence of moral disapproval does not indicate that there is no general causal connection (aversion ⟶ taboo); perhaps it's just a sign that we are here dealing with a particular kind of situation in which, by a very special logic of hypothesis testing, the aversion could not have produced the taboo. Let me explain.

Westermarck's aversion hypothesis asserts that there is an inborn sexual aversion that spontaneously develops between siblings, qua close childhood associates. The hypothesis faces an obvious difficulty. Namely, since as a matter of fact the aversion is usually correlated with the corresponding cultural taboo against sexual contacts between siblings, how do we know that the aversion is not simply a consequence of the taboo (rather than being an inborn, biologically mediated inhibition)? Well, fortunately for the aversion hypothesis, there are three extensively researched cases in which there is no taboo but the aversion still seems to be there.[9] These cases constitute critical empirical support for the aversion hypothesis. But, somewhat mischievously, these same cases undermine the expression hypothesis. To spell it out, on one hand, if the correlation between the aversion and taboo is occasionally broken, this is a welcome result for the aversion hypothesis because the aversion occurring without the taboo shows that the aversion stands by itself and is not just a side effect of the taboo. But on the other hand, the aversion occurring without the taboo is bad news for the expression hypothesis; it is weakening the hypothesis because what the hypothesis basically says is that, other things being equal, the aversion leads to the taboo. So, Westermarck's global theory about incest has two components, the aversion hypothesis and the expression hypothesis, that pull empirical evidence in the opposite directions. For this reason, his global theory is in a strange epistemological predicament in that, under the circumstances, it just cannot receive full empirical confirmation. It is a zero-sum game; what the theory gains by collecting evidence in favor of the aversion hypothesis it automatically loses on the other front because the very same empirical data chip away at the expression hypothesis.

Because of the zero-sum nature of Westermarck's argument, we should not be much impressed by the fact that in every single case that supports the existence of the biological aversion the taboo is missing. This is dictated by the logic of the situation.

Referring to Figure 6.3 and the two circles, the area of intersection represents the presence of both the aversion and the taboo. These cases are consistent with the expression hypothesis, but they cannot provide evidence for the aversion hypothesis. If all the data were in that area, the aversion hypothesis would be considerably weakened because then it would make sense to hypothesize that the aversion is produced by the taboo and is not an independent psychological phenomenon. However, the area where aversion occurs without the taboo sends an opposite epistemological message. The data points located in that section support the theory that the

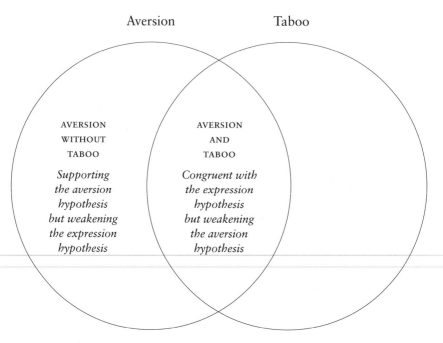

Aversion   Taboo

AVERSION
WITHOUT
TABOO

*Supporting
the aversion
hypothesis
but weakening
the expression
hypothesis*

AVERSION
AND
TABOO

*Congruent with
the expression
hypothesis
but weakening
the aversion
hypothesis*

FIGURE 6.3. The Expression Hypothesis and the Aversion Hypothesis:
An Epistemological Tension

aversion is an autonomous event and not a mere offshoot of the taboo. But
at the same time this break of connection between the aversion and the
taboo creates a problem for the expression hypothesis, which states that
the aversion regularly brings about the taboo.

So, there is nothing puzzling about the fact that all three cases supporting the aversion hypothesis speak against the expression hypothesis. Instead of being regarded as a consilience of independent cases strongly
pointing toward the probable falsity of the expression hypothesis, this fact
is better seen as just reflecting the logical peculiarity of Westermarck's theory. Although the two components of his theory, the aversion hypothesis
and the expression hypothesis, are perfectly compatible and mutually consistent, there is an epistemological tension between them in that, at least
at the present stage of theory testing, empirical evidence cannot support
the aversion hypothesis without raising some doubts about the expression
hypothesis.

For the purpose of illustration, let me give another example that comes
from an entirely different context but which, analogously, exhibits the same
kind of epistemological tension. Take the following hypothesis: "Many peo-

ple who approach the state of clinical death have the same strange experience: it appears to them that they perceive the world from a location outside their own bodies." Let's call this hypothesis NDE (near-death experience). Take now another hypothesis: "Usually, a near-death experience is immediately followed by death." Let's call this latter hypothesis DSA (death-soon-afterward).

It is quite clear that, from a logical point of view, NDE and DSA are fully consistent with one another. But here again, as in the case of Westermarck's two hypotheses, there is an epistemological strain. For, if the evidence in favor of DSA becomes too strong, then as a direct consequence of that, NDE will lose much of its empirical support. Namely, if the regularity postulated by DSA were perfect and exceptionless, and if, accordingly, all cases of NDE were immediately followed by death, there would be simply no one left to tell us about the existence of NDE! On the other hand, and this is the key point, if there happened to be some cases where, contrary to the general tendency expressed in DSA, the subjects of NDE survived and were able to report later about their experiences, it would be *wrong* to use this to attack DSA, along the lines reminiscent of Durham's argument in connection with Westermarck. That is, it would be wrong to say, "Of all the cases supporting NDE, not one shows evidence for DSA. Any one of these cases might be taken as exceptional, and therefore dismissed from concern. But it is hard to accept that argument for all of them."

True, all the cases supporting NDE undermine DSA, or at least weaken it to some extent. But this, in itself, should not be interpreted as a consilience of several independent instances, pointing to the probable falsity of DSA. Rather, the fact that all the cases corroborating NDE do clash with DSA is better regarded as just trivially following from the zero-sum logic of confirmation involving these two hypotheses. To put it differently, if from the outset it were quite open whether the data supporting NDE will happen to be in accord with DSA or not, and if it then turned out that all of them were aligning themselves in the opposition to DSA, this could indeed be reasonably taken as accumulation of important negative evidence against DSA. But this is not how things are. Actually, we know well in advance that any confirmation of NDE (i.e., an ex post facto first-person report about NDE) will inevitably be a counterexample to the regularity asserted in DSA ("NDE is immediately followed by death"). For this reason it is misconceived to ask, "Why is it that not just one or two cases confirming NDE undermine DSA, but all of them do?" The answer is, It could be no other way. The data speak here with one voice because of the epistemological peculiarity of the situation. This is not a probabilistic indication of anything.

Conclusion

I tried to show that Westermarck's theory about incest is not threatened by the two very general methodological objections directed against it.
But I hope that amid the critical tones, which dominate this chapter, a more
positive message will be recognized too. Namely, the epistemological
scrutiny of Westermarck's views that is here undertaken reveals a theoretical edifice of great complexity and conceptual sophistication. Although today the main efforts seem to be focused on the attempts to evaluate empirical evidence that will ultimately decide the conflict between the biological
account of incest taboos and its rivals, at those moments when we are
forced to stand back and inspect the logical structure of Westermarck's theory, we realize that some of its implications, interconnections, and "forking
paths" have yet to be fully explored.

NOTES

1. Edward O. Wilson, *Consilience* (New York: Knopf, 1998); Edward O.
Wilson, "Resuming the Enlightenment Quest," *Wilson Quarterly* (1998).

2. Richard Dawkins, "Opportunity Costs of Inbreeding," *Behavioral and
Brain Sciences*, vol. 6 (1983): pp. 105–6.

3. John Maynard Smith, "Constraints on Human Behavior," *Nature*, vol. 276
(1978): p. 121.

4. Neven Sesardic, "From Biological Inhibitions to Cultural Prohibitions:
How *Not* to Refute Edward Westermarck," *Biology and Philosophy*, vol. 13
(1998): p. 224.

5. See William H. Durham, Chapter 7 in this volume.

6. Steven Weinberg, "Against Philosophy," in *Dreams of a Final Theory*
(London: Hutchinson, 1993), p. 132.

7. Bernard Williams, "Evolution, Ethics, and the Representation Problem,"
in *Making Sense of Humanity and Other Philosophical Papers, 1982–1993*
(Cambridge: Cambridge University Press, 1995), pp. 105–6.

8. William H. Durham, *Coevolution* (Stanford, Calif.: Stanford University
Press, 1990), p. 323.

9. However, for serious methodological doubts about one of these cases
(the alleged evidence from Israeli kibbutzim), see Jonathan Hartung, "Review of
Incest: A Biosocial View," *American Journal of Physical Anthropology*, vol. 67,
no. 2 (1985): pp. 169–71.

7 Assessing the Gaps in Westermarck's Theory

William H. Durham

The last two decades of the twentieth century were just as kind to Edward Westermarck as the first two decades were harsh. The skepticism and dismissal that plagued Westermarck's incest theory in the wake of early critiques by Sigmund Freud and Sir James Frazer have given way to a recent groundswell of empirical validation and approbation. Today, Westermarck's theory is often held up as paradigmatic of current understandings of the relationship between genes and culture in human evolution. Frans de Waal, for example, writes that the Westermarck effect—that is, the absence of sexual interest between adults who were reared together as young children—serves as "a showcase" of new Darwinian approaches to human behavior.[1] Other authors claim that sibling incest avoidance not only vindicates Westermarck but also shows that "a tight and formal connection can be made between biological evolution and cultural change."[2] And in his recent treatise on the unity of knowledge, E. O. Wilson notes that Westermarck's argument is, simply, "the current explanation" of incest avoidance.[3] Outside observers of this changing tide could well be given the impression that Westermarck's theory has triumphed of late over all alternative explanations for the incest taboo and that we are dealing at the turn of the century with an open and shut case for Westermarck.

Such a conclusion would be grossly misleading. In the rush to vindicate Westermarck, touched off by careful documentation of the Westermarck effect (described by Patrick Bateson in Chapter 1 and Arthur Wolf in Chapter 4), it has proved easy to forget that Westermarck's is a complex theory containing *at least three* main hypotheses. Empirical support for one of the pieces of that theory, even if it is the crucial starting piece, is not the same as support for all three. For Westermarck to sweep into the new millennium with full validation, we would need equivalent evidence to support all three main hypotheses:

1. The *aversion* hypothesis, or "Westermarck effect"—namely, the proposition that "an innate aversion to sexual intercourse [develops] between persons living very closely together from early youth" (Westermarck 1891: 320).

2. The *adaptation* hypothesis—the proposal that the aversion is an evolutionary adaptation, shaped by natural selection specifically for the observed function of reducing inbreeding, and not some incidental by-product of the nervous system (and thus an "exaptation" in contemporary evolutionary parlance).

3. The *expression* hypothesis—the claim that the aversion directly causes the incest prohibitions of human societies; in Westermarck's words, "aversions which are generally felt readily lead to moral disapproval," and do so by "displaying" or "expressing" themselves "in custom and law as a prohibition of intercourse between near kin."[4]

There are probably many reasons why the first hypothesis on this list has received the lion's share of recent attention. For one thing, it flies in the face of the long-standing Freudian position on sibling attraction and has thus required thorough documentation even to budge the skeptics. As Arthur Wolf notes (in the Introduction to this volume), hypothesis one is thus a direct challenge to the received wisdom of earlier generations on the subject of incest. For another thing, the Westermarck effect lends itself fairly readily to empirical testing. A number of researchers (including Wolf, Joseph Shepher, and Justine McCabe) have investigated social settings offering some form of "natural experiment" in which cosocialized children are both reared together and allowed—indeed encouraged—to have sexual relations and/or marry.[5] Hypotheses two and three are much harder to get at in this way, and there has been much less concerted effort to do so. Understandably, number one carries the day.

But I remain concerned about hypotheses two and three. Before we usher in Westermarck's theory as the new crown prince of incest theory, and especially before we build from it toward a full-blown "biological science of ethics" (see Chapter 10), these two hypotheses seem to me to warrant the same kind of scrutiny given hypothesis one. Number two concerns me because the Westermarck effect is increasingly *assumed* to be an evolutionary adaptation, an assumption that renders it eligible for the "just so story" critique.[6] Its position is especially precarious because there are, at first glance anyway, other arguments—habituation among them (see Chapter 1)—to account for the deeroticization of cosocialized children. What is lacking is conclusive evidence to show that the aversion was specifically shaped over time by genetic selection for the function it now performs. While we await such evidence, it would be prudent not to rush the crown prince to the throne.

Parallel concerns apply to the expression hypothesis, but even more so.

For example, one could well argue that it doesn't really matter if Westermarck was right about the aversion being an evolutionary adaptation. As long as the nervous system delivers the effect, however it is achieved, Westermarck could still be correct in arguing that the aversion is the basis for the incest taboos of humanity. The crown would still be his. But the situation is not the same for the expression hypothesis. Westermarck could well be right about hypotheses one *and* two and still be fully wrong about what causes the incest taboos of humanity. To date, we have little more than mere arguments for the expression hypothesis. As rich with reason and logic as they may be, arguments are not the same as data. What is needed is empirical evidence that the internal, individual reaction of aversion (whether adaptation or not) "displays itself in custom and law as a prohibition to intercourse between near kin."[7] We need convincing demonstration that "the law" *exists because* it "expresses the general feelings of the community and punishes acts that shock them."[8] I argue that the crown should be held in reserve until these data are delivered.

In the remainder of this chapter, I would simply like to expand on these concerns about hypothesis three and the reasons they persuade me toward caution today, in the face of what otherwise looks like a headlong rush for Westermarck.

The Evidence Problem

In Westermarck's formulation, the incest taboos of humanity are a direct social manifestation of the aversion—they are the aversion as it "expresses itself in custom and law." The causal chain is straightforward and direct; the incest taboo culturally encodes the aversion people feel and it punishes others—at least it threatens to—if they carry out acts against which most members of the community are averse. To Westermarck, the aversion is thus the basis for the moral disapproval of incest within any given community; the reason people oppose and prohibit sexual acts between certain categories of kin is the discomfort they feel when others carry out those acts. In parallel to the "Westermarck effect," let me call this the "Westermarck process": the hypothesis that people actively condemn in others the sexual actions toward kin that arouse the aversion in them.

At first glance, one would think it ought to be fairly easy and straightforward to test the Westermarck process against empirical data. The questions seem simple enough: Do people actively oppose the sexual acts of others that make them feel aversive discomfort? Does a given community's disapproval of incest spring from these individual reactions? Does this aversion-based disapproval shape and sustain the incest taboos of humanity? In practice, however, tests have proved difficult; certainly there have not yet been enough efforts in this direction. There are, I think, only three things we can confidently say these days about testing the Westermarck process.

First, as Neven Sesardic has pointed out, not every aversion gives rise to a prohibition, so there must be something special about the Westermarckian aversion.[9] In fact, human societies appear to harbor quite a number of aversions that remain unmatched by prohibitions: the aversion to eating insects alive;[10] the aversion to drinking fresh milk among nondairying peoples;[11] a widespread aversion to bitter, poisonous substances; a purported aversion to left-handers by right-handers;[12] and so on. If it is truly valid, the Westermarck process must have something special about it that does not pertain to these other aversions. Wolf (1995 and Chapter 4 of this volume), building on Westermarck's (1906–8) own arguments about "moral ideas", suggests that what is special here is the combination of disinterestedness, apparent impartiality, and the "flavor of generality." As Wolf argues,

It is in the nature of inhibitions of this kind to arouse moral disapproval. The inhibition is activated as an aversion when a person expresses sexual interest in an object belonging to the same class as those objects toward which most people are indifferent; the pain [or discomfort] produced by the thought of sexual relations with a sexless object is expressed as disapproval of the person responsible for raising the possibility; and this disapproval is accepted as moral because, conferring no obvious advantage, it appears to be disinterested, and, being the reaction of the majority of the community, it qualifies as impartial and general.[13]

In other words, inbreeding aversion is a special kind of aversion, and the Westermarck process is but one example of a more general process behind the social emergence of moral ideas. It is a lovely and provocative idea, especially if it succeeded in getting everyone to believe that the expression hypothesis needs no testing because it is merely a special case of an accepted general phenomenon. The problem is that the argument is based on a Westermarck publication from 1906–7 that shows clear signs of "having been modeled after his [own] earlier theory about incest prohibition."[14] In short, Westermarck's whole line of reasoning behind "the origin and development of moral ideas" still remains to be tested, in general *and* in the case of the incest taboo.

A second point to be made is that cross-cultural tests of the Westermarck process have consistently failed to produce much evidence. Westermarck was himself the first to try with a test of a corollary argument: namely, the argument that the range of prohibited relatives in a given society should closely parallel the group of relatives typically reared together from an early age. He claimed,

Facts show that the extent to which relatives are not allowed to intermarry is nearly connected [tightly correlated] with their close living together. Generally speaking, the prohibited degrees are extended much further . . . [in societies where people] live, not in separate [nuclear] families, but in large households or communities, all the members of which dwell in close contact with each other. . . . On the other hand, where families live more separately, such extensive prohibitions to close intermarrying do not generally exist.[15]

In an extension of this reasoning, Westermarck even argued that coresidence and "local relationships" explain why incest prohibitions are "very often one-sided, applying more extensively either to the relations on the father's side or to those on the mother's, according as descent is reckoned through men or women."[16] He offered the general conclusion that "an abundance of ethnographical facts . . . prove that it is not, in the first place, by degrees of consanguinity, but by the close living together that prohibitory laws against intermarriage are determined."[17]

It was a noble thought, a logical corollary of the Westermarck process, and a valiant attempt at empirical validation. But it was also wrong. To Westermarck's credit, later editions of his book set the record straight; he dropped the preceding conclusion and noted in its place that taboos "frequently refer to the marriage of kindred alone," *not* to persons reared in propinquity.[18] Still in hopes of reconciling this finding with his theory, these later editions made explicit a couple of additional steps in Westermarck's chain of logic. First, he argued that the aversion to sex with close childhood associates is "interpreted" by people as a feeling against sex with kin. This "transition" occurs, says Westermarck, because the aversion is supposedly one of "an immense group of facts which, though ultimately depending upon close living together, have been interpreted in terms of kinship."[19] Second, he added to this transition his famous "law of association,"[20] which is variously seen by scholars today either as a brilliant, symbolic extension or as one of the more interesting and problematic fudge factors in all of social science. By this law, the "feelings of intimacy and kinship" from close childhood association are transferable to other, nonintimate acquaintances. As a result, incest taboos, "though in the first place associated with kinship because near relatives normally live together, have come to include relatives who do not live together."[21] The transfer is facilitated, says Westermarck, by symbols—especially names—that have "come to stand" for consanguinity. This certainly does help to generalize the Westermarck effect beyond the limits of childhood association. But it also makes the argument much more difficult to test. It becomes something of an open question whether the resulting prediction—that "the extent of the prohibited degrees is closely associated with social intimacy, whether combined with actual living together or not"[22]—is even falsifiable.

Meanwhile, a couple of other attempts at cross-cultural tests of the Westermarck process have also failed to produce empirical support. The first of these was an ambitious study of first-cousin marriage practices in a global sample of some 800 populations.[23] Reviewed in more detail elsewhere,[24] Ember's study looked for evidence in this sample of societies that childhood association among cousins was correlated with their cultural prohibition as marriage partners. His measures of childhood association were, first, community endogamy (that is, rules about marrying within one's natal community) and, second, community size, on the reasonable argument that cousins are more likely to be reared in close association when they inhabit the same village, especially if it is small (i.e., less than 400 members). Granted that these measures of propinquity were, at best, indirect, the results were still challenging to Westermarck. First-cousin marriages were actually prohibited with *greater* frequency where cousins are routinely precluded from childhood intimacy by community exogamy. Moreover, the smaller endogamous communities (presumably conducive to greater social intimacy) failed to show a significantly higher proportion of cousin prohibitions than did endogamous communities of a larger size. Ember concluded that the childhood association of cousins has little value in predicting their prohibition as marriage partners.

In a much less ambitious study in the 1980s, I attempted a retest of Ember's findings in a world probability sample of sixty societies[25]—a sample with a higher proportion of independent cases (less of "Galton's problem") than Ember's much larger sample.[26] I also used a measure of the kinship extension of incest taboos that is far more refined than Ember's "yes" or "no" on cousin marriage prohibition.[27] To make a long story short, despite my best efforts, these modifications did little to change the conclusions reached by Ember. I, too, found that locally endogamous populations, with presumably *more* opportunity for close relatives to share intimate childhood association, do tend also to be "more accepting of close relatives as sexual partners than do their locally exogamous counterparts."[28] And like Ember, I also found that small endogamous communities (in which more relatives might be expected to grow up in close childhood association) tend to have the *least* extensive incest taboos. At the same time, ethnographic details in both studies, Ember's and mine, make it abundantly clear that there generally is strong moral disapproval of incest within the societies studied, albeit variably expressed. There is just no evidence from *either* analysis that the moral disapproval of these communities is based on Westermarck's aversion. By the same token, both studies did find modest support for an alternative source of moral community in the study populations, an alternative source that will be discussed further below. In short, existing cross-cultural studies, including Westermarck's own efforts, provide little or no support for the hypothetical Westermarck process.

The third and final point to be made about testing the Westermarck process concerns Sesardic's critique (Chapter 6) of my earlier discussion of three special cases—minor marriages in Taiwan, intra-Kibbutzim marriages in Israel, and patrilateral parallel cousins in Lebanon—from which we get the best data for Westermarck's aversion.[29] Upon reviewing each case, I noted that while all three support "Westermarck's argument about intimacy and aversion, not one shows evidence for a moral disapproval expressive of that emotion."[30] Sesardic takes me to task for making so much of this point, since we already know that the aversion could not have produced a taboo in these cases, or the cases wouldn't exist. He has a good point; it is only possible for us to see and document the aversion in cases where there is no taboo against the aversion-producing relationships in the first place (if there were a taboo, as this older logic goes, any aversion might instead be a product of the taboo). Says Sesardic, if you've chosen these cases because the relationships studied are free of taboos, then of course you can't then expect the same cases to show evidence for the Westermarck process! They were chosen precisely because they have no taboo for the relationships concerned.

I accept Sesardic's criticism and agree that the cases can't be expected to show both a taboo-free aversion and an aversion-based incest taboo. That said, the cases remain interesting to this discussion not simply because the absence of a taboo makes them exceptional. They also stand out because the aversive couples are actually socially desired marriage partners. To my mind, it is the *positive social valence* to these unions that makes them interesting, not the (logically necessary) absence of a negative valence. In the case of minor marriage in Taiwan, it is clear from Wolf's accounts that the parents of cosocialized couples have a lot invested and eagerly desire the marriages to work, even to the point of brandishing canes outside bedroom doors of reluctant newlyweds. In the case of age-mates in the Kibbutzim, says Shepher, the marriages found to be aversive were actually *"preferred* by parents and other members of the kibbutz."[31] And in the case of Lebanon, McCabe makes it clear that "one of the most salient features of marriage in the Arab Middle East is the preference for a man to marry his patrilateral parallel cousin," despite the aversion the two feel.[32] In all three cases, the aversion is accompanied by some preference for the aversive union. Sesardic might well come back with the reply that, yes, they almost *had to be* preferred forms of marriage in order for them to reach frequencies adequate for scholars to be able to measure aversion effects. But look at what else that tells us. The existence of social preferences for marriage between aversive individuals tells us that the hypothesized Westermarck process can be overridden by other, everyday social phenomena. In at least three well-documented cases, any aversion-based disapproval is overwhelmed and turned right around, such that aversive unions are coupled with community moral *ap-*

proval. Now, Wolf might well come back and add that this is not all that surprising, because aversions developing out of childhood association may be "typically experienced in kinship terms" (from the Introduction to this volume). In both Taiwan and Israel, the childhood associates are not kin and thus may be exempt from a disapproval experienced in kinship terms. But what about Lebanon? In Lebanon, the aversive individuals are certainly kin (first cousins). What is more, in Lebanon, the turnaround is nothing new. For hundreds of years, maybe even thousands, the aversive unions of close cosocialized kin have met with enduring community approval. To quote Westermarck once again, the aversion here simply does *not* "display itself in custom and law as a prohibition to intercourse between near kin." To my mind, all three exceptional cases, and especially the Lebanese example, raise the question of the relative strength of aversion-based disapproval compared to other ordinary social forces in human communities. If routinely overridden in Lebanon, are we to suppose it is strong enough in all other times and places to deliver the nearly ubiquitous incest taboos of humanity? I remain to be convinced.

The Burton Problem

But there is one other problem, perhaps more serious than evidence issues, suggesting that it may still be premature to hand Westermarck the incest crown. This is simply the problem of an alternative hypothesis that both (1) links the Westermarck effect with incest taboos in another way, and (2) garners enough empirical support of its own to be difficult to dismiss out of hand. Certainly the alternative has more supporting cases than there are with clear evidence for the Westermarck process. I call it the "Burton problem" because psychiatrist Roger Burton was, I believe, the first author to formulate the argument parallel to Westermarck's theory, beginning with the aversion but running right on through to moral communities that sustain incest taboos primarily for other reasons.[33] I confess at the outset that I was struck many years ago by both the simplicity of the argument and its base of empirical support. I was persuaded at that time to carry out a related analysis of the Burton argument, and thus I do not pretend to be nonpartisan in this debate.[34] However, I also do not claim that the Burton problem is the only conceivable alternative explanation to Westermarck's. Other alternatives may well surface in the years ahead, perhaps more worthy than any of our current "short list" of theories. That fact should surely be borne in mind before we rush the crown out to Westermarck, or to his competition for that matter.

The Burton problem, then, is basically the problem (from Westermarck's vantage point) of an alternative explanation linking aversion and the incest taboos of humanity. The alternative occurred to Burton in the course of a cross-cultural study on a different topic, resistance to temptation, in which he was given "the impression that the most common reason given in both primitive and modern societies for the incest taboo is that [inbreeding] produces bad stock."[35] It occurred to Burton that there may well be something more than coincidence or superstition to these reports and that the local observation and experience of genetic effects of inbreeding could well affect the scope (i.e., the extension beyond the nuclear family) and persistence of human incest taboos.

To Burton's credit, he realized that it was precisely in societies where Westermarck's aversion kept childhood associates at bay that the reduced frequency of inbred unions would allow for the evolutionary accumulation of deleterious genes in a population.[36] In a sense, Westermarck's starting hypothesis is what set up Burton for a rival theory of four parts:

1. Westermarck's aversion would guarantee that inbreeding was generally rare in local gene pools but commonly deleterious and visible when it occurred. Just how deleterious and visible remains a matter of some debate (see Chapter 2, for example), but Burton reasonably assumed that the additional mortality and morbidity from inbreeding among close relatives (first cousins and closer, say) would commonly be enough for people to notice their damaging effects over the sweep of time.

2. According to Burton, local peoples would then, over the years, come to recognize the abnormal, undesirable consequences of inbred unions, despite the absence of a Western scientific understanding of their genetic causes. Eventually, they would come to see the effects both as detrimental and as distinctive from other afflictions caused by accident or contagion in their population.

3. Once recognized, people would commonly attribute these special harmful consequences to the displeasure of the supernatural—a warning, essentially, that the mating that produced the deformity (or deformities) was a transgression of normal order (maintained by the aversion) and was being punished by supernatural agents. People would see inbreeding as divinely punished, which would no doubt contribute to the special affect it is so often accorded.

4. Fear of additional retribution from the supernatural and fear that multiplication of these infractions and their consequences could bring harm to all—at the hands of nature or hostile enemies, for example, if not at the hands of deities themselves—would promote moral disapproval of the responsible act (or acts) within the group and would prompt continuing prohibition of that which displeased the supernatural.

As formulated by Burton, the theory had much to recommend it. It gave Westermarck's aversion an appropriate and prominent role; it fit nicely with emerging genetic theory for inbreeding effects, while also not assuming that local people had any knowledge of Western genetics; it accommodated a wide range of local beliefs, assuming only that supernaturals of some kind had the power to deform fetuses and/or prevent newborns from thriving; it carries a strong built-in motivation for individual compliance (a fearful threat of supernatural sanction); and contra Bateson (Chapter 1) it does propel a conviction that others should be prevented from violation, or sanctioned after the fact, lest the community suffer further disasters at the hand of supernatural forces. Moreover, Burton's argument immediately explains why Westermarck had, himself, come upon so many, many "folk theories" of inbreeding effects. In fact, Westermarck's original list of folk theories—five pages of cases compiled innocently in support of his claim that inbreeding could have caused the aversion to evolve by natural selection[37]—remains in my view the single best compilation of data for the Burton theory.[38]

But even with all these positive features, I was struck early on by several limitations to Burton's original formulation. The argument needed expanding beyond simply the genetic effects of inbreeding to include *any* reproductive impairment or negative "fitness effects" stemming from sexual relations with kin, including the psychological consequences of sexual trauma and any social stigma affecting inbred families.[39] In the right circumstances, these additional sources of reproductive impairment could well contribute to the locally recognized consequences of inbreeding, as indeed could the Westermarck effect itself (through lowered fertility, etc.). But also needed was a broadening of Burton's notion of attribution and supernatural punishment. Rather than assume, as Burton did, that locals would always think "the gods" were punishing close inbreeding via its reproductive effects, for indeed some societies might well have gods lacking in such powers, it made more sense simply to leave as an open matter the cultural interpretation of inbreeding effects, as long it is locally meaningful, widely convincing, and negatively valued. The resulting hypotheses, (1) that local peoples recognize at least some adverse consequences to inbreeding, and (2) that they culturally interpret them as harmful and potentially threatening to everyone, still constitute a form of "bad stock" argument. But they are more inclusive and realistic than Burton's original suggestions. And they are also more readily incorporated into an "optimal outbreeding" framework, in which any such perceived "inbreeding costs" might be balanced against prevailing "outbreeding costs" in the eyes of the locals.[40]

Empirical Support for Burton's Theory

Once strengthened by these and other modifications, the Burton argument seemed to warrant testing in an unbiased sample and so, in an earlier study, I tested its predictions against data from the same equiprobability sample of sixty world societies mentioned earlier.[41] The sample is admittedly tiny for any attempt at global generalization, but that fact made it possible to carefully scrutinize, with the help of Stanford student assistants, the large (though partial) ethnographic database available for each society of the sample—the Human Relations Area Files.[42] To make another long story short, I found that the HRAF data for forty-two of the sixty societies contained at least one reported consequence to incestuous congress, including social and supernatural sanctions. But more important, I found that twenty-three of these forty-two cases—54.8 percent—featured some recognition of inbreeding effects. Moreover, in a number of those twenty-three cases, including the Tikopia of Polynesia and Eastern Toradja of Sulawesi, there was enough ethnographic material to confirm that both recognition and cultural interpretation were clearly at work.

Consider the case of the Eastern Toradja, a population of dry rice cultivators who inhabit the mountainous interior of Sulawesi and are distant linguistic relatives of the Trobrianders discussed by Hill Gates (Chapter 8). As informants explained to Dutch observers Albert Kruyt and Nicolaus Adriani (missionary and linguist respectively, whose combined work in the region ranged from 1890 to the 1940s), the Toradja incest taboo stems from a specific physical incompatibility:

It can happen that a man and a woman have physical characteristics (*oea ngkoro*) that come into conflict with each other when they marry. The harmful influence of this will manifest itself in the children born from such as marriage: they will be weak, sickly, or idiotic and quickly die. Therefore a marriage with a person too closely related is considered unsuitable (*bare'e raposioea*, literally, not corresponding in physical characteristics); their children are then not healthy and do not live long.[43]

In other words, Toradja informants cite the harmful consequences in offspring as the reason inbreeding is prohibited, consequences they interpret as evidence of a "conflict of physical characteristics." Not only do these consequences closely resemble the expected genetic effects, but there is also (an accurate) recognition that the effects are less likely at greater kinship distances among consanguines. Adriani and Kruyt found that many Toradjans "object to a marriage between cousins out of fear that such a union will remain childless; or that the children resulting from it will not have viability, [or] will be idiotic or crippled."[44] In contrast, when the blood relationship is more distant, marriages can still be arranged provided that certain rites of expiation are also observed, in which case (also accurately) children are likely to be healthy and thrive.

The Toradja case thus seems a good general fit with the expanded Burton argument. Close inbreeding is "unsuitable" because of its reproductive consequences. It is not in the first instance that the gods are angry; rather, it is a simple physical incompatibility of characteristics. Later on, the gods may indeed manifest displeasure in the form of widespread drought or damaging rains, according to Adriani and Kruyt,[45] but the telltale signal of the problem, the reason inbreeding is considered "unsuitable," is incompatibility. (An alternative hypothesis that suggests itself—that this "bad stock" theory represents simply a diffusion of Dutch understandings of inbreeding—is readily rejected by clear evidence that indigenous terms for "incompatible characteristics" and related phenomena greatly predate the arrival of the Dutch in 1905.) So one might reasonably ask, Why does incest matter to others in a village, beyond the immediate affected parents and close kin? Here again the Toradjan case seems exemplary even when there is no drought; prior to pacification in the 1900s, there was almost incessant concern over the economic and military vitality of one's village given the prevalence of headhunting in the region. Notes Robert Lagace, "Until European contact, the Eastern Toradja lived in a state of essential village autonomy and semiperpetual hostility with each other. . . . Intervillage raiding served to provide scalps necessary for many rituals and to pacify spirits (*anitu*) who would otherwise feed on the Eastern Toradja."[46] In this environment, a village rose or fell with the strength of its able-bodied warriors and workers.

But there is still one further observation by Toradjan chiefs that is, to me, the most convincing part of all in regard to the Burton argument. According to Adriani and Kruyt (1951, 2: 56–57) again,

There are many stories about men and women who supposedly [had sex with] a tree spirit [*bela*]. There are women who say they are visited each night by a *bela*. Many become pregnant from such intercourse. . . . The children who are said to result from [sex] with a spirit are described as "light of skin with blue eyes and white hair." This refers to albino children, who are rarely found among the Toradja. But other characteristics of children whose father is supposed to have been a spirit are mentioned. One of them was hairy, "like a monkey;" another had the nose and eyebrows of a monkey. . . . So-called spirit children do not live long, because their fathers [the *bela*, are said to] take them in order to bring them up themselves.[47]

Significantly, the authors report that "more than one Toradja chief assured us that women claimed to have been made pregnant by a tree spirit when their condition was the result of intercourse with a member of her kin group whom she was not permitted to marry." Not only does this local interpretation provide explanation for the early death of inbred children, but the data also show that *chiefs take physical deformity of offspring as diagnostic of illicit incest*. They, at least, are not fooled by the alibi.

In summary, data from the Toradja and twenty-two other cases in the global sample offer striking if partial support for a different theory linking Westermarck's aversion to the prohibition against incest. The cases are fully complementary with Westermarck's own list of folk theories mentioned above, with the additional advantage of not being gleaned from the literature to make the particular point. (Indeed, they are joined by a further example from his separate study of marriage in Morocco: "It was also the opinion of the ancient Arabs that the children of marriages between relatives are weakly and lean. Thus a poet [wrote] ' . . . the seed of relations brings forth feeble fruit.'")[48] But with only 54.8 percent of the pertinent societies recognizing inbreeding effects, these findings are best seen as simply provocative, far from a full-blown empirical verification of Burton's alternative explanation. Still, they should at least give us pause. They show that non-Western human populations do *commonly* recognize the deleterious effects of inbreeding and interpret them as a negative sign or warning that such unions, for a variety of reasons, invite these effects. At the same time, bear in mind that the other, nonconfirming cases of the "Sixty Cultures" sample warrant careful interpretation; they specifically do *not* mean that these other peoples fail to recognize and culturally interpret inbreeding effects. All they mean is that there is *no report* of such phenomena in the (admittedly spotty) HRAF database of this study. On balance, then, these data support the conclusion that many societies—including Western ones—have incest taboos that are derived fully or in part from the observation of inbreeding effects.[49]

This conclusion has, in my view, a number of implications for contemporary scholarship in the general area of incest theory. First, it is definitely a bit early to suggest that incest taboos "*had nothing to do* with society not wanting to look after the half-witted children of inbreeding, since in many cases they had no idea that inbreeding was the cause."[50] If we may trust an admittedly tiny probability sample, then the conclusion here implies that in more than half of all societies known to anthropology incest taboos have *much* to do with people seeking to avoid sickly, weak, and half-witted children. No doubt some analysts will say that "more than half of all societies known to anthropology" is still disappointing, given the near universality of the incest taboo (on universality, see Chapter 5). To this argument, it must again be noted that Burton's theory has far more supporting evidence than exists for the Westermarck process.

Second, the conclusion here implies that it is time for social science to move beyond a lingering prejudice about the observational powers of non-

Western peoples. Such a prejudice was bluntly articulated, for example, in 1948 by anthropologist Leslie White, who remained unconvinced by even the genetic arguments of his time:

But suppose that inbreeding did produce inferior offspring, are we to suppose that ignorant, magic-ridden savages could have established this correlation without rather refined statistical techniques? How could they have isolated the factor of inbreeding from numerous others such as genetics [sic], nutrition, illnesses of mother and infant, etc., without some sort of medical criteria and measurements—even though crude— and without even the rudiments of statistics.[51]

I believe the data presented here suggest that White's "ignorant, magic-ridden savages" did frequently figure it out, even without the advantages of Western medicine and statistics. That said, let me emphasize that I have no quarrel with the suggestion by Gates (Chapter 8) and others to the effect that "hypertrophied cultural logic" about good, healthy offspring and the conditions that cause them, rather than the experience of the negative consequences of inbreeding, may have initially provoked the cultural interpretations of proper and improper mating. I am not sure it matters which came first really, and I doubt we'll ever know for sure. All that matters is that the connection be made *somehow*, convincing people that "although X, Y, and Z promote healthy offspring, inbreeding does not." What matters in the history of incest taboos, I submit, is this linkage, no matter which way around it is culturally constructed.

Finally, on a related point, I am not suggesting that inbreeding beliefs are free of superstition or of "added" or "piled on" consequences. As Westermarck was himself first to point out, all manner of things—"epidemics, earthquakes, sterility of women, plants or animals, or other calamities"— get heaped onto the pile of purported incest consequences in diverse societies, adding in the local purview to its feared, harmful effects.[52] Data available for the sixty societies described here are no exception. However, the data summarized above do call into question two of Westermarck's corollaries to that observation. First, they question his assertion that, among non-Western peoples, the incest taboos, "especially those related to the nearest relatives, are so strictly observed that no genuine knowledge could possibly be based on the few cases in which they are transgressed."[53] In many cases of the "Sixty Cultures" study described here, Westermarck is directly contravened on the topic, sometimes as explicitly as this:

In a conversation which I had with [an informant] on the subject of incest he brought up the matter of supernatural sanction as follows. "Some brothers and sisters," he said, "who have the one father but different mothers, will join together, will embrace each other. When such brother and sister have begotten their children, these keep on dying, dying, dying, and the labor is of wailing, wailing, wailing for the children who do not exist long, but die off . . . " That the offspring of such incestuous marriages do die in this way he was prepared to support by definite evidence, like other informants. "The observation of it in this land is finished," he said, meaning that there were cases to hand, known to everyone, which formed the empirical basis for the common opinion.[54]

Second, these data make me think Westermarck doth protest too much: "Even if we had a right to make such an assumption, even if savage men everywhere had discerned that children born of marriage between closely related persons are not so sound and vigorous as others, we could certainly not be sure that they everywhere would have allowed this knowledge to check their passions."[55] The issue here is not whether knowledge checks passions, although I am sure that makes a lovely debate, or as Gates (Chapter 8) so aptly points out, over "who wins," reason or passion. The argument is over the taboo itself, a cultural product, and the way it is shaped by *both* passion *and* reason. Despite Westermarck's protestations and those of some supporters since, there remains little or no basis for rejecting the Burton hypothesis, particularly in its expanded form. It remains possible that Westermarck's aversion can be causally related to the incest taboos of human societies in ways that bypass Westermarck's process. It seems to me that this implication alone calls for more serious attention to the actual reasons that close inbreeding is condemned in human populations.

Conclusion

Ironically, after presenting these and related ideas at the Stanford conference, I actually came away wondering if we have been emphasizing the wrong question all along about incest taboos. The question we seem to have been asking is, What explains the nearly ubiquitous incest taboos of human societies? It occurs to me now, and is at least worth putting forward for further reflection and discussion, that this question really parses into two separable questions: (1) How did the incest taboos of humanity originate? and (2) What forces sustain and shape incest taboos in the diverse human populations of the ethnographic record? The first of these questions is very difficult, if not impossible, to answer with our current analytical tools and data. It may well be that we simply can't get at origins, and thus any attempt to "pit" Westermarck versus Burton on this question may be completely in vain. I question whether we'll ever, in the foreseeable future, know enough to fairly test alternative hypotheses about origin.

The second question, in contrast, seems like a much more tractable topic. It, too, might be separated into two distinct questions, namely, What process (or processes) or force (or forces) are most important in sustaining incest prohibitions in human populations? And what processes operate to shape existing taboos, governing their extensions and limits as well as their sanctions? The Westermarck theory is certainly one good candidate for providing answers to these questions. However, if I have been successful in these remarks, the point will have been made that it is not alone in the candidate pool. Still to be eliminated, or perhaps somehow incorporated, is an interesting rival hypothesis—if nothing else a foil to today's more popular Westermarck arguments. The rival holds that in most, or maybe all, human populations of the ethnographic record, incest taboos are sustained and shaped primarily by a local moral disapproval that is expressed in terms of (what Western science calls) "inbreeding effects" as they are locally and meaningfully interpreted.

Could the incest taboos also have originated in this way, by early recognition of inbreeding effects and appropriate cultural interpretations in a variety of ancestral populations? Yes, I think they could have, especially if one regards human cultural history as a frequently branching tree (such that not all populations had to recognize and interpret inbreeding effects independently and from scratch). But I also think it will be a long time, if ever, before we can say anything like that with confidence. Could the incest taboos have originated some other way, such as by Westermarck's arguments, and then later had recognition and interpretations of inbreeding effects added on? Yes, this is also quite possible and could well bring to human history certain advantages to both theories. For example, the automatic "innate action" of Westermarck's process could later have been joined by the more cultural, interpretive action of the Burton theory, with the advantage I should think of a stronger, more locally meaningful cultural reinforcement of the taboos. Yes, I think a variety of historical scenarios are possible, many of them potentially integrating Burton and Westermarck components. But I also think it will be difficult to get clarity and resolution when dealing with questions of origin. Instead, the take-away points I hope to have made in this chapter are two: first, the case for Westermarck is by no means open and shut as we begin the new millennium, and second, whenever it occurred, the recognition of inbreeding effects and their local cultural interpretation by human observers added powerful cultural evolutionary forces to the sustaining and shaping of incest taboos. They added nothing less than symbolism and meaning to the moral emotions.

NOTES

1. F. B. M. de Waal, "The End of Nature Versus Nurture," *Scientific American*, vol. 281 (1999), pp. 94–99.

2. M. Ruse and E. O. Wilson, "Moral philosophy as applied science," *Philosophy*, vol. 61 (1986), p. 184. See also C. J. Lumsden and E. O. Wilson, *Genes, Mind, and Culture: The Coevolutionary Process* (Cambridge: Harvard University Press, 1981.

3. E. O. Wilson, *Consilience: The Unity of Knowledge* (New York: Alfred Knopf, 1998), p. 173.

4. Edward Westermarck, *The History of Human Marriage*, 5th ed., vol. 2 (New York: Allerton, 1922), pp. 198 and 193, respectively.

5. See review in William Durham, *Coevolution: Genes, Culture, and Human Diversity* (Stanford, Calif.: Stanford University Press, 1991), chapter 3.

6. See, for example, S. J. Gould, "Sociobiology: The art of storytelling," *New Scientist*, vol. 80 (1978), pp. 530–33; S. J. Gould and R. C. Lewontin, "The spandrels of San Marco and the Panglossian paradigm: A critique of the adaptationist programme," *Proceedings of the Royal Society of London, Series B*, vol. 205 (1979), pp. 581–98.

7. Westermarck, *The History of Human Marriage*, 5th ed., vol. 2, p. 193.

8. Ibid., p. 204.

9. Neven Sesardic, "How *not* to refute Westermarck," *Biology and Philosophy*, vol. 13 (1998), pp. 413–26.

10. Ibid., though not in all societies—see P. Menzel and F. D'Aluisio, *Man Eating Bugs: The Art and Science of Eating Insects* (Berkeley, Calif.: Ten Speed Press, 1998).

11. Durham, *Coevolution*, chapter 5.

12. Patrick Bateson and Paul Martin, *Design for a Life: How Behavior and Personality Develop* (New York: Simon and Schuster, 2000), pp. 216–17.

13. Arthur P. Wolf, *Sexual Attraction and Childhood Association: A Chinese Brief for Edward Westermarck* (Stanford, Calif.: Stanford University Press, 1995), p. 512.

14. Durham, *Coevolution*, p. 323.

15. Edward Westermarck, *The History of Human Marriage*, 1st ed. (London: Macmillan, 1891), pp. 324–27.

16. Ibid., pp. 329–30.

17. Ibid., p. 321.

18. Westermarck, *The History of Human Marriage*, 5th ed., vol. 2, p. 207.

19. Ibid., p. 205.

20. Ibid., p. 216.

21. Ibid., p. 214.

22. Ibid.

23. Melvin Ember, "On the origin and extension of the incest taboo," *Behavior Science Research*, vol. 10 (1975), pp. 249–81.

24. Durham, *Coevolution*, pp. 341–44.

25. See Robert O. Lagace, ed., *Sixty Cultures: A Guide to the HRAF Probability Sample Files, Part A* (New Haven, Conn.: Human Relations Area Files, 1977).

26. Durham, *Coevolution*, chapter 6.

27. For details see ibid., pp. 344–57.

28. Ibid., p. 353.

29. Ibid., pp. 310–14.

30. Ibid., p. 323.

31. Joseph Shepher, *Incest: The Biosocial View* (New York: Academic Press, 1983), p. 60.

32. Justin McCabe, "FBD marriage: Further support for the Westermarck hypothesis of the incest taboo," *American Anthropologist*, vol. 85 (1983), p. 50.

33. Roger Burton, "Folk theory and the incest taboo," *Ethos*, vol. 1 (1973), pp. 504–16.

34. Durham, *Coevolution*, pp. 324ff.

35. Burton, "Folk theory and the incest taboo," p. 505.

36. For discussion of the genetic theory behind this point, see W. M. Shields, *Philopatry, Inbreeding, and the Evolution of Sex* (Albany, N.Y.: State University of New York Press, 1983), chapters 3 and 4.

37. Westermarck, *The History of Human Marriage*, 5th ed., vol. 2, pp. 170–74.

38. Durham, *Coevolution*, p. 324.

39. With regards to sexual trauma, see D. Wilner, "Definition and violation: Incest and incest taboo, *Man* (n.s.), vol. 18 (1983), pp. 134–59; and with regards to social stigma, see S. Yokoyama, "Social selection and evolution of human diseases," *American Journal of Physical Anthropology*, vol. 62 (1983), pp. 61–66.

40. See Patrick Bateson, "Optimal outbreeding," in *Mate Choice*, ed. Patrick Bateson (Cambridge: Cambridge University Press, 1983).

41. For details see Durham, *Coevolution*, pp. 329ff.

42. See Lagace, ed., *Sixty Cultures*.

43. N. Adriani and A. Kruyt, *The Bare'e-speaking Toradja of Central Celebes (The East Toradja)*, vol. 2 (Amsterdam: N. V. Noord-Hollandsche Uitgevers Maatschappij, 1951). English translation in Human Relations Area Files, Section OG11, "Toradja."

44. Ibid., p. 284.

45. Ibid., p. 271.

46. See Lagace, *Sixty Cultures*, p. 411.

47. Adriani and Kruyt, *The Bare'e-speaking Toradja*, pp. 56–57.

48. Edward Westermarck, *Marriage Ceremonies in Morocco* (London, Macmillan, 1914), p. 55.

49. See, for example, Jack Goody on "The Letter of Gregory" behind the limits of the Catholic prohibition. Jack Goody, *The Development of the Family and Marriage in Europe* (Cambridge: Cambridge University Press, 1983), pp. 35–36.

50. Bateson and Martin, *Design for a Life*, p. 217, emphasis added.

51. Leslie A. White, "The definition and prohibtion of incest," *American Anthropologist* vol. 40 (1948), p. 418.

52. Westermarck, *The History of Human Marriage*, 5th ed., vol. 2, p. 178.

53. Ibid.

54. Raymond Firth, *We, The Tikopia: A Sociological Study of Kinship in Primitive Polynesia*, reprint edition, abridged by the author (Stanford, Calif.: Stanford University Press, 1983), pp. 288–89.

55. Westermarck, *The History of Human Marriage*, 5th ed., vol. 2, p. 178.

8 Refining the Incest Taboo

WITH CONSIDERABLE HELP FROM
BRONISLAW MALINOWSKI

Hill Gates

. . .

Ikatupwo'i inala: "Mtage luguta?"
She asks her mother: "Indeed my brother?"
Kawalaga: "O latugwa boge inagowasi!"
Her speech: "O my children already they are mad!"
. . .
Ilikwo dabela, iseyemwo.
She unties her fiber skirt, she puts it down.
Ivayayri namwadu. . . . Iloki luleta.
She follows the shore naked. . . . She goes to her brother.
. . .
Ibokavili . . . iyousi, ikanarise wala obwarita.
She chases, she takes hold, they lie down right in the sea.
Ikanukwenusi, ikammaynagwasi, ivino'asi.
They lie, they go to shore, they finish.
. . .
Gala ikamkwamsi, gala imomomsi, u'ula ikarigasi.
They neither eat nor drink, and so they die.

—From the Trobriand Myth of the Origin of a Love Magic
Strong Enough to Incite Incest (Bronislaw Malinowski,
The Sexual Life of Savages in North-Western Melanesia)

The *incest taboo* is a phrase that delivers an emotional and semantic double blow. The two words, each problematic enough alone, together befog the analytic impulse with a messy mix of biological, gothic literary, and arcane religious implications. My task here is to decompose that complexity, demystify some of its historically accrued sensationalism, and ground it empirically—without denying its genuine experiential explosiveness. What do we mean, and what have we meant, by the *incest taboo*?

To answer, I briefly explore the embrace by culture-oriented twentieth-century anthropologists of Sigmund Freud's Oedipus complex and its odd alliance with contract theory. At present, our accumulating knowledge of human biology aligns better with evidence from an earlier phase in ethnography, epitomized by the work of Bronislaw Malinowski. His account of Trobriand incest taboos emphasized local sociocultural influences on a common human nature and thus anticipates the late-twentieth-century conclusions of Arthur Wolf. Edward Westermarck, Malinowski, and Wolf would agree that incest taboos are rationalized expressions of innate distaste, with the core aversion locally embroidered to fit cultural context.

To further refine our understanding of incest taboos, we may consider how a universal sexual aversion within the nuclear family has been harnessed to social use in different political-economic clusterings of incest taboos. Here, as illustration, I isolate brother-sister incest as a possible factor in cultural evolution.[1]

In *Totem and Taboo*, Sigmund Freud did two things. He wrapped the term *incest taboo* in his era's uneasy religiosity and Lamarckian biology, and he incautiously accepted an obsolescent and dismissive view of the intellectual achievements of "primitive man." For Freud, the taboo was a prohibition invented in a remote evolutionary Dreamtime, when men "forbade themselves" sex with their mothers and sisters. Necessary for the emergence of civilization or culture, and absorbed as an inherent part of our human nature, the taboo was recapitulated in each individual as the price of sexual and social maturity. Freud thus linked two currently fascinating topics—human evolutionary origins and family conflict—while legitimating open discussion of individual sexual development. The rapid adoption of Oedipal explanation into Western popular thought probably owed as much to ethnographic ignorance and prejudice as to the delightful liberation of talking trash at dinner tables.

The curious connections Freud drew in *Totem and Taboo* immediately attracted support among anthropologists. This is perhaps because of a convenient intellectual slippage: between Freud's grand vision of humans constructing full humanity—person by person, people by people—through acts of will, and the sociology of knowledge that vivified post–World War I Western European thought.[2] Throughout the twentieth century, emphasis on the socially constructed character of human nature has dominated liberal-to-progressive social analysis. The need for such emphasis to counter rationales for customary race and gender biases will be obvious to readers of this volume.

Anglophone anthropologists especially committed themselves to such liberal positions, finding Freud, for all his increasingly obvious ethnographic naïveté, a charismatic ally. They ignored the inherent biologism of his position and recast his incest taboo as an eternally reinvented social contract limiting sexual conflict and binding vulnerable families into defensive kinship networks—as E. B. Tylor and Sir James Frazer had framed it previously. In the words of Claude Lévi-Strauss, "we are of the opinion that the prohibition of incest provides sufficient guarantee that a network of alliances, resulting in all other respects from free choices, will not compromise social cohesion."[3] Leslie White, Marvin Harris, Maurice Godelier, and many others would later concur.

According to this mainstream position, evolving humans eschewed the convenience (and perhaps desirability) of mating with close kin in order to force the formation of wider social networks through exogamy. Especially when combined with the Freudian premise of nuclear family attraction, this conquest of nature gave rise to "civilization" or "culture"—a new, human domain of experience that transcended all previous mammalian constraints.

In the twentieth century, most anthropologists viewed the willed passage via the incest taboo from animal to human as the very essence of culture in its general evolutionary sense. Once across the evolutionary bridge of renouncing incest, humans became free, socially plastic agents in a myriad of specific cultures—localized, contingent, historical, all capable of transcending our animal behavioral inheritance.

The confusion in anthropological discourse that arises from the conflation of two distinct uses of the term *culture* would be hard to overestimate. In an essay titled "Evolution: Specific and General," Marshall Sahlins discussed both biological and cultural evolution in these terms: "The distinction has long existed in the literature of evolutionary anthropology. E. B. Tylor, . . . (1871), laid out the study of cultural evolution both "stage by stage" as well as "along its many lines." Sahlins continues: "General cultural evolution . . . is passage from less to greater energy transformation, lower to higher levels of integration, and less to greater all-round adaptability. Specific evolution is the phylogenetic, ramifying, historic passage of culture along its many lines, the adaptive modifications of particular cultures."[4] The later Marshall Sahlins and similarly relativist anthropologists now repudiate "general cultural evolution" to describe the differences between egalitarian, ranked, and state societies as ways of life or "stages." But none would disagree, I suppose, that the transition to fully modern humans exhibiting the capacity for symbolic thinking—"culture"—represents such a general evolutionary move. It is helpful, then, to retain Sahlins's distinction between culture general and culture specific. It flags the difference between culture as a uniquely developed human capacity to transmit information and cultures as localized and historically contingent products of this capacity.

If biological factors cannot explain cultural difference—as they surely cannot—it has also been assumed that biology can teach us nothing about our species as a whole. It is deemed useless (or racist or sexist) to investigate possible genetic and developmental constraints or unities of a species nature.

Further, if a common human nature operated on our varied cultures, we should see a set of human universals to class with the incest taboo. But, culturologists emphasize, most nineteenth-century cultural universals have been successfully deconstructed. Feminists have purged scientific thought of many common fantasies; students of comparative religion and race have dissolved others. Even kinship, marriage, and the nuclear family, important contenders for universality in our species, have been vigorously rejected in some quarters.[5] That the incest taboo is a willed override of nature by culture, however, is generally agreed—often in the most amazingly muddled terms.[6]

This consensus is increasingly untenable. While human aversion to incest surely has consequences for social organization and individual maturation, to take these consequences as causes is to commit one of the most frequently and most legitimately condemned sins of functionalist interpretation. The line that contract theory draws between "us"—humans—and "them"—the rest of nature—inevitably seems less sharp to thinkers who know how many of their genes they share with a fruit fly.

After a short twentieth century of Freudian obfuscation and overemphasis on culture, research on inbreeding, incest, and the incest taboo is achieving a new clarity and reliability. Rearguard defenses of their exclusively cultural origin are still mounted. By the twenty-first century, however, they can be supported only by stubborn citation of out-of-date material.[7] Evidence from nonhuman behavior, especially among primates, as well as from large and well-documented human communities has returned debate—as this volume demonstrates—to the serious consideration of evolutionary factors. The insight of Darwinian anthropologist Edward Westermarck, long hidden under the Freudian bushel, now has empirical support (see Chapter 4). A focus on human nature by the anti-Freudian anthropologist Bronislaw Malinowski, derided since the 1930s as psychologistic naïveté, now appears prescient.

Evaluating Ethnographic Evidence

Anthropological generalizations can be sustained only when they meet the test of evidence got by two very different methods, each flawed in a different way. The ethnographic case study must be thorough, and it will always require the most exacting contextualization: Where? When? As interpreted by whom? For what audience? Cross-cultural comparisons are vexed instead by the "unit problem": What groups of people constitute distinctive sociocultural entities? In our calculations with these units, how should we weigh two thousand Tikopia against a billion Chinese? The difficulties are obvious, generating argument that sometimes descends to academic brawling.[8] In the United States especially, the influence of British structural-functionalists waned rapidly after World War II, and the Marxist and ecological anthropology of the 1960s and 1970s flourished even more briefly. Almost by default, much twentieth-century ethnography has been shaped by research programs with a strong culturalist bent. By the end of the century, even word-wizard Clifford Geertz deplored this tendency while comparing work before the cultural turn with that of his own sorcerer's apprentices: "All the human sciences are promiscuous, inconstant, and ill-defined, but cultural anthropology abuses the privilege."[9]

Fortunately, earlier, more holistic, and more comparative anthropologists have left us bodies of work sufficiently rich in detail and wide-angled in viewpoint to extricate us from this intellectual cul-de-sac. We turn to what is arguably the best single ethnographic description of an incest taboo complex, embedded in Bronislaw Malinowski's investigations of Trobriand Island life during World War I, and then to insights from quantified, comparative studies.

Most of Malinowski's observations of and conclusions about Trobriand society and culture were published in five detailed and respectfully written books.[10] His many other publications included materials about the Trobrianders and their near neighbors, especially *Magic, Science, and Religion*; his methodology is made transparent in an unvarnished account of his Trobriand fieldwork—*A Diary in the Strict Sense of the Term*—that appeared posthumously.[11] In *Sex and Repression in Savage Society*, he repudiated a transiently felt Freudianism, admitting himself to having been one of "many fools" deeply impressed by the promise of psychoanalysis and its Oedipal core.[12] These he replaced with a full and nuanced discussion of Trobriand incest that emphasized local sociocultural influences on a common human nature.

This mass of material has survived anticolonialist and feminist critique better than most ethnography of its time. J. P. Singh Uberoi and Marshall Sahlins expanded on but did not dispute Malinowski's interpretation of the region's political economy as one where vigorous staple markets were elaborately constrained by ritual gift exchange among local leaders.[13] Annette Weiner researched Trobriand women's work in the 1970s with greater gender-attentiveness, adding much to our understanding of the overall system.[14] She might easily have assumed what E. P. Thompson called the "enormous condescension of the young" toward her predecessor. Instead, she rejected the argument "that ethnographic writing can never be more than a kind of fictional account of an author's experiences. Although Malinowski and I were in the Trobriands at vastly different historical moments and there also are many areas in which our analyses differ, a large part of what we learned in the field was similar."[15]

Weiner has relieved us of another academic anxiety by commenting on the effects of colonialism during Malinowski's research. As Trobrianders sometimes told Malinowski, times had changed with the coming of foreign administration, legal codes, and plantation labor.[16] Fifty years later, Weiner noted the surprising stability of Trobriand practice.[17]

For myself, I should like to see a thoughtful revision of the kinship/gender/inheritance/status nexus that often puzzled Malinowski as a contradiction between matriliny and father-right. Since his time, anthropologists have developed more sophisticated interpretations of ranking, lineality, and gender roles. David Labby has given us an exemplary reconception of Oceanian social process that can illuminate Trobriand experience;[18] Karen Sacks's Africanist perspective is also provocative.[19] True to form, however, Malinowski himself collected enough data to substantiate such models, so very different from his own.

Perhaps the most valuable quality of Malinowski's ethnography is its recognition of both the importance and the striking limitations of local texts. Getting out of the armchair, off the guesthouse verandah, and beyond the simplicity of symbolic systems was an intellectual as well as a methodological breakthrough. Insisting on attention to enacted as well as to enunciated meaning, Malinowski observed:

[Indigenous] statements show us the polished surface of custom which is invariably presented to the inquisitive stranger; direct knowledge of native life reveals the underlying strata of human conduct, moulded, it is true, by the rigid surface of custom, but still more deeply influenced by the smouldering fires of human nature. The smoothness and uniformity, which the mere verbal statements suggest as the only shape of human conduct, disappears with a better knowledge of cultural reality.[20]

He imputes no deceit to his respondents. A Trobriander "simply does what any self-respecting and conventional member of a well-ordered society would do"—that is, he summarizes reality with norms and ideals;[21] "native peoples" have as much difficulty as anyone in describing their reality to outsiders. "For in actual life rules are never entirely conformed to, and it remains, as the most difficult but indispensable part of the ethnographer's work, to ascertain the extent and mechanism of the deviations."[22] Ethnographers who rely principally on direct interrogation get bad data. "Such material [leads] to the anthropological doctrine of the impeccability of native races, of their immanent legality, and inherent and automatic subservience to custom."[23] From Malinowski, we learned that only researchers with local experience, full command of the relevant language, and his own generous comprehension of our common humanity could hope to capture and transmit the reality of "the," or even "an" incest taboo. One can only imagine his salty response to the notion that culture might differentially construct, and thus negate, such common human emotions as attachment and loss.

Incest Among Trobriand Islanders

One among many peoples who inhabited the horticultural, fishing, artisan, and trading region of northeast New Guinea and its offshore islands, Trobrianders operated a complex economy in which both women and men played necessary roles.[24] Matrilineal in clan descent and property inheritance, the key adult female/adult male dyad was not wife and husband, but (as in matriliny generally) the brother/sister pair. Jointly inheriting various properties from their mother and mother's brother (a clan brother if not a uterine one), a sister and brother were jointly responsible for rearing and socializing the next generation, and for passing to them land rights, canoe shares, and other properties inherited through a line of women. At harvest, a man (and his wife or wives) carried the better half of his harvested yams, the staple food, to his sister's household for its use.[25] In return, women produced quantities of banana leaf bundles without which their matrilineage brothers could not maintain political competitiveness and meet kin obligations.[26] A mother's brother fed his sister's children—his heirs— while she herself worked to guarantee their future rights.

A boy's mother's brother was expected to teach him rituals that "belonged" to his matrilineage. A boy's biological father—usually his mother's husband—often sentimentally but illicitly shared such secrets and other property with the sons he reared and loved, stealing them from his proper heirs—his sister's children. A boy's "mother's brother assumed a gradually increasing authority over him, requiring his services, helping him in some things, granting or withholding his permission to carry out certain actions; while the father's authority and counsel become less and less important."[27] Trobriand brothers and sisters were life partners with shared interest in their young heirs, much common property, and continuing mutual attachment.

At the same time, any hint of sexual connection between them was sharply sanctioned.

Brother and sister thus grow up in a strange sort of domestic proximity: in close contact, and yet without any personal or intimate communication; near to each other in space, near by rules of kinship and common interest; and yet, as regards personality, always hidden and mysterious. They must not even look at each other, they must never exchange any light [i.e., sexual] remarks, never share their feelings and ideas. And as age advances and the other sex becomes more and more associated with love-making, the brother and sister taboo becomes increasingly stringent.[28]

Even in childhood, expression of affection between them was punished.[29] Like Freud, Trobrianders appear to have conflated nonsexual attachment among close childhood associates with sexually based bonding; Malinowski, as we shall see, did not.

Malinowski described the prohibition of brother-sister sexual relations as "the supreme taboo of the Trobriander," "the prototype of all that is ethically wrong and horrible to the native. It is the first moral rule seriously impressed in the individual's life, and the only one which is enforced to the full by all the machinery of social and moral sanctions."[30] Because human beings came into the world as a brother-sister pair between whom sex was forbidden, Trobrianders normatively described conception as a result of connection between a woman and her ancestral spirits. The absence of a male contribution to pregnancy was a logical necessity in the Trobriand myth of origin, and a position that many Trobrianders held to be factual.[31] From all directions, their culture instructed Trobrianders to cherish an intimate but completely nonsexual bond between sister and brother.

A sibling without his or her opposite number was disadvantaged in such a domestic economy. Among Trobrianders, as among other lineally organized peoples, the extension of kinship throughout the clan mitigated such difficulties. Matrilineally related boys and girls were classificatory brothers and sisters to each other, falling under the incest taboo.

Among Trobrianders, Malinowski expanded,

Sexual intercourse and marriage are not allowed within the same totemic clan. They are more emphatically forbidden within a sub-division of the clan, common membership in which means real kinship. And the taboo is stricter yet between two people who can trace a common descent genealogically. Yet the natives have only one word, *suvasova*, to designate all these degrees of exogamous taboo. Also, in legal and formal fiction, the natives would maintain that all exogamous taboos, whether of clan, sub-clan or proven kinship, were equally binding.[32]

Yet a man's marriage with his father's sister's daughter—or even a father's sister—was common, at least in some subclans.[33] Prohibition and moral disapproval of clan mating shaded off as the matrilineal connection became more remote; incestuous affairs and even marriages occurred on the fringes.[34]

On the surface we have one word, *suvasova*, one clan kinship, one punishment, one sense of right and wrong. In reality we have the distinction between marriage and mere intercourse, between clan and sub-clan . . . , between genealogical kinship and mere community of sub-clan, between the own sister and the classificatory sisters. We have also to distinguish between direct enforcement by public opinion, and by supernatural sanctions, neither of which works in a simple or infallible manner.[35]

The idea of son-mother incest provoked real distaste, and Malinowski explored this topic with considerable care.

The nearest female of the previous generation, the mother, is also surrounded by a taboo, which is coloured, however, by a somewhat different emotional reaction [from that elicited by brother-sister incest]. Incest with her is regarded with real horror, but both the mechanism by which this taboo is brought home and the way in which it is regarded are essentially distinct from the brother-sister taboo. The mother stands in a close bodily relation to her child in its earliest years, and from this position she recedes, though only gradually, as he grows up. As we know, weaning takes place late, and children, both male and female, are allowed to cuddle in their mother's arms and to embrace her whenever they like.[36]

Sons need not conceal their sexual activities from their mothers (or fathers); and

since normal erotic impulses find an easy outlet [with peers], tenderness towards the mother and bodily attachment to her are naturally drained of their stronger sensuous elements. Incestuous inclinations towards the mother are regarded as highly reprehensible, as unnatural and immoral, but there is not the same feeling of horror and fear as towards brother and sister incest. When speaking with the natives of maternal incest, the inquirer finds neither the rigid suspense nor the emotional reactions which are always evoked by any allusion to brother and sister relations. They would discuss the possibility without being shocked, but it was clear that they regarded incest with the mother as almost impossible. I would not affirm that such incest has never occurred, but certainly I have obtained no concrete data, and the very fact that no case survives in memory or in tradition shows that the natives take relatively little interest in it.[37]

Because they are members of the same household and because his daughter is his wife's nearest kinswoman, intercourse between father and daughter or stepdaughter is taboo.[38] "We do not sleep with [a daughter]," said Trobrianders, "because [the father] fondles [and] takes [her] into his arms" as a child.[39] It occurred,[40] but reality was complicated by a kin terminology that lumped father and father's sister's sons—including classificatory ones. If a true father-daughter relationship came to light, it might shame the man into suicide, but it was not categorized as *suvasova*, "exogamy breaking," and did not result in disease.[41] The possibility that father-daughter (or any) incest might be punished by imperfect or sickly offspring is nowhere mentioned; a father who is not a biological progenitor fits poorly into any argument from indigenous assessment of inbreeding depression.

When discussing taboos in general, Malinowski distinguished among "the genuine taboos with supernatural sanction, the clear prohibitions without supernatural sanction, and prohibitions of acts which must not be done because they are [so transparently] shameful, disgusting, or else dangerous."[42]

The breach of the *suvasova* taboo entails a "supernatural" penalty: infestation by an insect "spontaneously generated by the actual breach of exogamy," which covers the skin with sores and produces pains and discomfort throughout the body.[43] Trobrianders can wax eloquent on this not necessarily fatal ailment: "As the natives put it: 'We find maggots in a corpse. How do they come? *Ivagi wala*—it just makes them. In the same way the insect is made in the body of the . . . exogamy breaker. This insect wriggles round like a small snake; it goes round and round; it makes the eyes swollen, the face swollen, the belly swollen, like in [dropsy].'"[44]

This etiology calls into question Malinowski's use of the term *supernatural*, which we may assume Trobrianders would not use to describe maggots coming from a corpse any more than my own premicroscope ancestors did for the same phenomenon. Throughout his career, Malinowski struggled to escape the inherent ethnocentrism of the Western concepts magic, religion, and science. Usually careful about the difficulty of drawing a boundary between "natural" and "supernatural" in Trobriand ontology, here he fails to resolve a significant hermeneutic difficulty. Incest dermatitis is a "supernatural" but not an agentive punishment; it is in the nature of things.[45]

Conveniently, the illness "entailed" by exogamy-breaking has "a perfectly well established remedy against any pathological consequences of this trespass, a remedy considered practically infallible, if properly executed. . . . [This is] a system of magic consisting of spells and rites performed over water, herbs, and stones, which when correctly carried out, is completely efficient in undoing the bad results of clan incest."[46]

For Trobrianders, as among ourselves, Malinowski denied a "slavish adherence to tradition," even about incest, insisting on the full and self-contradictory humanity of all people.[47]

[He] was told that if a man came by chance upon his sister and her sweetheart while they were making love, all three would have to commit *lo'u* (suicide by jumping from a coco-nut palm). This is obviously an exaggeration which expresses the ideal and not the reality: if such a mishap occurred the brother would most likely pretend to himself, and to them, that he had had seen nothing, and would discreetly disappear.[48]

What human sanctions *were* invoked against the incestuous?

The only socially acknowledged "brother/sister" incest case occurring during his fieldwork is described in *Crime and Custom in Savage Society*.[49] A boy had an affair with his mother's sister's daughter, a fact "known and generally disapproved of." Her previous boyfriend, jealous, accused him in public of incest, shaming him beyond endurance. The following morning, the male exogamy-breaker dressed and ornamented himself, climbed a coconut tree, and bade farewell to his clansmen. Accusing his enemy of driving him to his death, he called on his clansmen to avenge him, leaped, and died. Taking up his cause, his relatives wounded the rival and held the customary mortuary ceremonies for him. His cousin-sweetheart, we learn elsewhere, married, lived happily with her husband, and can be seen in the frontispiece of *Sexual Life of Savages*.[50]

The Malasi, a clan with very highly ranked subgroups, were unusual, associated in fact and story with intraclan incest. Malinowski made no mention of social sanctions or suicide among them. I will return to the unorthodox Malasi toward the end of this chapter.

Does such loose fit between norm and practice in incest sanctions simply indicate that Trobrianders never directly punished breakers of social norms? No. Malinowski collected accounts of a man speared for sorcery, "a few cases" of killing of adulterers in flagrante delicto. Insulting a chief risked this punishment.[51]

Self-punishment through suicide did occur; Malinowski heard of several people who had jumped from palms and several others who had poisoned themselves, although none for reasons of incest.[52] In a "well-known" case, a man caught having sex with an animal was expected to commit suicide, but he did not. "The culprit . . . has lived down his shame. He leads a happy existence in Sinaketa, where I had the pleasure of meeting him, and having a long conversation with him."[53]

From his careful investigation of a subject of great interest to both himself and his informants, I can only conclude that even first-degree incest rarely resulted in serious perceivable harm to the islanders. Although incest evoked strong negative emotions, they knew it to be less dangerous than adultery or lèse-majesté. Malinowski's observations led him to believe that sexual attraction within the nuclear family was unusual; that when it occurred, it had only minor and occasional consequences; that sexual attraction was common, and marriage not uncommon, among relatives at a few removes; and that culturally defined incest at some remove was seen as troubling rather than reprehensible, unlikely to result in retribution from any quarter. The Trobrianders' incest taboo had been shaped by the local political economy around a "natural" core of what Patrick Bateson has identified as optimal outbreeding.[54]

Reason Versus Passion

Malinowski ended his career still emphasizing that human beings, by nature, had needs that social organization and culture must to some degree accommodate. Everything he wrote urges us to believe that anyone could correctly interpret behavior in other societies if we understood how their local peculiarities intersected with our shared human nature. By the time Raymond Firth published *We, the Tikopia* in 1936, such "psychologizing" —necessarily biological at base—had largely been abandoned by anthropologists, however. Even though Firth's volume had a foreword from Malinowski, Firth himself was convinced that the ultimate explanation for incest taboos was not biological but sociological.

The incest-exogamy attitudes may not be reducible to a simple formula. I am prepared to see it shown that the incest situation varies according to the social structure
of each community, that it has little to do with the prevention of sex relations as
such, but that its real correlation is to be found in the maintenance of institutional
forms in the society as a whole, and of the specific interest of groups in particular. . . .
    . . . where interest of rank or property steps in, the incest prohibition is likely to
melt away.[55]

In the late twentieth century, even this relatively grounded sociological approach was being swamped by a powerful idealist trend among anthropologists. Annette Weiner chided Malinowski because "his functionalist theories
obscured the subtleties and the significance of social action. His interest was
in the cause and effect of certain actions and activities rather than in the cultural meanings that Trobrianders give." Having completely dismissed the biological premise of Malinowski's frequent allusion to causal emotions, she
assumed that Malinowski was arguing for an absolute rationalism.[56]

The incest taboo stood through the twentieth century as the last of anthropology's agreed-on human universals, one thought by Tylor, Firth, Lévi-
Strauss, and many others to require no biological basis. Anthropologists
turned to the social contract outlined by Firth or to complete cultural constructionism as favored by Weiner at the expense of natural sentiments.
Scottish Enlightenment Passion with its Humean acceptance of affect disap-
peared behind French Enlightenment Reason.

In the light of late-twentieth-century findings, however, the disjunction
between these two positions grows increasingly uncertain. Each depends,
ultimately, on a recognized or unrecognized base in our unique human biology. To explain incest taboos, contractarians like Lévi-Strauss assume
evolved, species-specific cognition; Westermarck and Malinowski assume
the importance of our kind's unusually rich emotions.

Contractarians explain that the incest taboo is invented when, perceiv-
ing the advantages of out-marriage, men exchange their close female kin to
make allies. (In our enlightened times, contractualists might grant women
the rationality to see such advantages for themselves. Let us assume so,
sidestepping a potential red herring.) From such a position, the role of cognition in a rational allocation of resources is thrown into especial relief. In
the deep past, we became human through exercising a new intellectual capacity that our nonhuman ancestors did not have.

A more convincing, noncontractarian argument from human rationality
has been developed by William H. Durham. Durham approaches anthropological questions through a dialectical relationship—a coevolution—between biology and culture. Accepting the existence of incest-aversive emotion, he asks, "What is the relationship of incest taboos to the phenomenon
of inbreeding depression? and What is their relationship to the sexual aver-

sion [that is now known to develop] from childhood association?"[57] He answers that we have coevolved cultural incest prohibitions along with a genetically based aversion because we perceive a relationship between inbreeding and inbreeding depression. Durham takes particular note of the appearance in norms and folklore of a connection between incest and such visible consequences of inbreeding as weak and defective offspring.

Proposing a cost-benefit model derived from optimal outbreeding theory, he assumes that "the individual-level costs and benefits of inbreeding and outbreeding can be measured on a scale of individual inclusive fitness."[58] With this model, he predicts that incest taboos, as memetic self-replicators, "will have high cultural fitness in any population where there are substantial reproductive costs to inbreeding."[59] "Although the recognition and interpretation of reproductive impairment are not logically required for taboos to attain high [fitness], they are nevertheless predicted to provide a major force of cultural selection in most populations."[60]

Durham then applies his model to a sixty-culture sample from the Human Relations Area Files. He concludes from these comparisons that incest taboos are genetically adaptive and that they "may be viewed as memes that have been kept 'alive' in cultural pools largely through the positive cultural evaluation of their consequences."[61] Rational actors perceive the inbreeding depression that may result from incest and construct the negative cultural evaluation of inbreeding that we call an incest taboo.[62]

Durham's argument appears to require "that value-guided cultural selection has been the principal mechanism of change in the evolutionary descent of incest taboos."[63] Necessarily, if not sufficiently, cognition and rationality more than aversive affect are the aspects of human nature that shape incest taboos. Like Adam and Eve and Lévi-Strauss, we achieve full humanity through choice.

Edward Westermarck's hypothesis and Arthur Wolf's tests of it are, as this volume illustrates, very differently based. In Wolf's account of the Westermarck effect, the core incest aversion is rooted in affect that unfolds as Bowlbian attachment behavior in normal infants; a similar conclusion was reached by the psychiatrist Mark Erickson.[64] Attachment, essential to infant survival, precludes the later unfolding of mature sexual attraction for the object of attachment and even renders most people averse "when the act is thought of." A developmentally nuanced sexual aversion for former caregivers is a human universal, as is the tendency to generalize individual repugnance to moral disapproval. Incest taboos thus are rationalized expressions of innate distaste, with the core aversion locally embroidered to fit cultural context.

Comparing Durham's with Wolf's approach to this problem illustrates an important emerging tendency in evolutionary anthropology. We see the formation of two camps: those like Durham, who emphasize intelligent perception, rational choice, the partly intentional construction of culture, and a fairly clear line between ourselves and all other species; and those like Wolf, who stress urgent affect, contextualized post facto rationalization of an aversion, the interactions of sociality, and the strong persistence of primate traits. Once put so plainly, this sharp dichotomy is instantly suspect. Remembering the mess that generations of "nature versus nurture" thinkers made of social theory, one hopes that a fruitless polarization of reason and passion can be avoided in *this* century.

Brother-Sister Incest as a Mechanism of Cultural Evolution

Each locally elaborated incest taboo may thus be a distinct creation of reasoned minds working in concert with a unifying heritage of resistant emotion. The aversion is not "the incest taboo." Societies construct many incest taboos; each will be distinct, the product of a line of dizzying contingency.

Still, human experience is not a chaos of infinite possibility.

Beyond strong hints of the aversion itself, what elements of well-described incest taboos can be seen to cluster in informative ways? Do incest taboos and taboo-breakings point us toward some analytical middle ground between an exaggerated universality and a promiscuous particularity of explanation? Can we learn something broader about the nature and evolution of societies by focusing on this vivid commonality?

Well-grounded efforts toward the evolutionary categorization of societies are older than Lewis Henry Morgan and returned during the 1960s and 1970s. Jack Goody pioneered a comparative structural analysis of incest prohibition by pairing it with bans on adultery.[65] Marxist and ecological thinking revisited cultural evolution and infused anthropology with resurgent respect for what Marvin Harris saw as the materials from which all our cultures are constructed: "guts, sex, energy, wind, rain, and other palpable and ordinary phenomena."[66]

Now, as materialism returns to anthropological theory, Stevan Harrell's highly original attempt to find order in the spectrum of human families builds especially on the work of Morton Fried and Jack Goody and resonates with many of Malinowski's and Westermarck's concerns.[67] Like most theorists who keep one foot firmly on material conditions, Harrell clusters societies by political-economic criteria. He is clear on the direction of influence between family form and wider institutions: "On the whole, it is change in the larger social organization that brings about family change rather than the other way round. . . . [Conditions for organizing families] are set by the nature of the social system in which the family system is embedded, though the family system may play a small part in determining the nature of that larger system.[68]

Harrell's powerful typology of social systems specifies the contexts within which human families take their relatively limited forms. Harrell pays close attention to kinship in ranked societies: those with marked differentiation in public positions that is nevertheless insufficient to restrict access to subsistence/reproduction goods, and where those defined as chiefs are exempted from ordinary subsistence labor.[69] Harrell's emphasis on ranking suggests an approach to cultural difference in rationalizations of incest aversion, especially when the general tendency of aversion to brother-sister marriage is reversed to favor such marriages.

Brother-sister mating appears as by far the commonest, if not the only, form of mating within the nuclear family that has been linked with specific political-economic form. In an archaeological and ethnohistorical study of the chiefly Calusa of early colonial southwest Florida, John Goggin and William Sturtevant list most of the known societies that accepted brother-sister marriage; thirty-six of the forty-two distinguish either by rank (in Fried's terminology) or by class (in state societies).[70]

Some state societies are also known for sibling marriage, especially in ruling houses. It is suggestive that no well-developed bureaucratic state is known to have urged its royalty to marry incestuously, with one exception: Roman Egypt.[71] Apart from Roman Egypt, Goggin and Sturtevant list only small, newly formed, or recently barbarian-conquered states (Bali, Cambodia, the Hittite polity, Incan Andes, Java, Korea, Sinhala, and Thailand). Most of the rest are complex chiefdoms like Hawai'i, structurally perched to tip over into the class rigidities of state-ness. Strong chiefdoms, like small, aristocratic kingdoms were heavily dependent for power-creation on what Clifford Geertz called "theater state" tactics: flashy monumentalism, spectacular public ritual (often with human sacrifice), and the lavishly detailed apotheosis of rulers.[72] This repertory of cultural flamboyance accords well with the shock value of royal incest. Brother-kings married sister-queens, or brother-chiefs their sister-chiefs, flouting a prohibition commoners will have felt to be natural. Lords and ladies of the earth flaunted superhuman invulnerability, constructing auras of power by haughty taboo-breaking. They did so especially when they had not yet invented the administrative and communications systems to enforce their rule by less colorful means. Such families had the power to override aversion and enforce such marriages even on unwilling couples.

Polynesian peoples developed a number of high-chiefdom/almost-state societies in which brother-sister marriage for the high-ranking was politically salient. Computer simulation of Tongan chiefly marriage appears to confirm that brother-sister marriage influenced social structure by significantly increasing

all indicators of stratification, including the status variability that existed among chiefly lines and the status differences that developed among chiefly individuals. The results of the simulation suggested further that marriage rules might have been necessary to the system in order for stratification to occur because, without the presence of endogamous customs, the trajectories of chiefly statuses converged.[73]

Where conditions for ranking were auspicious, a family that risked inverting taboo and ignoring aversion set in train a snowballing status improvement for its descendants and an evolutionary leap in social complexity for its society. Somewhere in the history of all early states we might expect to find royal brother-sister incest lurking.

Return to the Trobriands

Malinowski too had something to teach us about the possible role of brother-sister incest in the evolution of political-economic forms. Throughout his work, the great ethnographer called Trobriand leaders "chiefs"—*guya'u*—stressing the ranking between chiefly and nonchiefly subclans. Writing before the term *chief* stabilized at Fried's definition, Malinowski followed the discourse of the day that labeled any "native leader" as a chief. He viewed a man like To'uluwa of Omarakana as holding a position equivalent, or nearly so, to those of the chiefs of the smaller Polynesian societies—perhaps the Tikopia leader, the Ariki Kafika.

This usage was disputed by J. P. Singh Uberoi, who concluded that Trobriand *guya'u* more closely resembled the Melanesian "big man," a nonheritable position dependent on continuous political-economic competition.[74] It can easily be shown that Oceanic peoples invented many local leadership forms that fit uneasily into the typologies of 1960s evolutionary approaches. Trobriand *guya'u* seem to me to be one of them. What Malinowski saw in the 1920s may well have been a system of leadership that lucky and clever men could tip from classical Melanesian nonheritable power to something much more like enduring Polynesian chiefly status for a descent group—at least for a time. Malinowski emphasizes (though he does not identify) one mechanism in such a process: men's inclination to dispossess their proper heirs—their sisters' sons—of inheritances in favor of their wives' sons, whom they had helped rear and come to love.

In such a labile situation (made somewhat more so, surely, by the colonial setting), were high-status Trobrianders experimenting with brother-sister incest as another such mechanism, a shockingly visible index of power? The possibility is *almost* documented.

The Malasi clan claims to rank higher than all others, and among them, the members of subclan Tabalu take first place: "they are the real chiefs, acknowledged to be of supreme rank, not by the Trobriands only, but by the adjoining areas as well."[75] They prefer to marry within other Malasi subclans, often with male paternal first cousins, a relationship "regarded somewhat askance" by most Trobrianders.[76] Interestingly, in view of Small's findings about the polarizing effect of in-marriage, the Malasi also includes a subclan whose members "were the most despised in the entire region (despite their superior industry and artistry), forced into endogamy with their own kind."[77] We are reminded of the castelike and maritally endogamous sociopolitical relations in Micronesia between high-ranking Yapese and the lowly, industrious Ulithi, who supplied Yap with fine handicrafts in return for essential raw materials.[78]

Malinowski also tells us,

Of the four [clans], the Malasi have the reputation of being the most persistent exogamy-breakers and committers of incest. All the incestuous marriages on record have happened within this clan; and I was told that this was not an accident but that only the Malasi and no other clan will tolerate such marriages. The myth of incest . . . is associated with the Malasi, and so also is the magic of love and the magic to frustrate incest disease.[79]

One of the previous paramount chiefs, Purayasi, was known to have lived with his sister; and another one, Numakala, is also strongly suspected by history of this felony. They, of course, belonged to the Malasi clan; and there can be no doubt that with them, as with so many other dynasties and famous rulers, the feeling of power, of being above the law, served as a shield from the usual penalties. And, as historical figures, they and their doings would not so easily lapse into oblivion as in the case of commoners.[80]

When Stevan Harrell defines historically derived political-economic types, he argues for the necessity of families to adapt to social and ecological structures over which family strategies have little leverage. Erickson has pointed out (personal communication, March 8, 2001) that using the theater-state tactic of brother-sister incest, a small, powerful family might well "strengthen its political hold over the larger group of families." Under the right circumstances, agency operates, with change in family practice triggering political-economic transformation.

We may now conclude that although incest taboos vary widely, they are necessarily responsive to an evolutionarily driven, biologically based aversion for associates of the first few years of life, who are usually members of the nuclear family. At considerable human cost, the aversion may be overridden.[81] Local adaptation to material constraints and cultural repertory will result in considerable variation in incest taboos. Taboos vary greatly (though not infinitely) in both their cultural conformity to the aversion and in their occasional dramatic inversion of it in normative brother-sister marriage. Such marriages occur almost exclusively in societies where ranking is sufficiently developed to exempt high-status people from most manual labor, as in chiefdoms, or to give them class rights over society's basic means of production, as in states. And they are unlikely to occur where state-building had created sufficient mundane power to relieve the ruling class of most of its supernatural aura. Under conditions not fully mapped out, but surely recurrent in human history, our innate alertness to the emotional complexity of incest can be turned to precise political ends, until something more dependable comes along.

NOTES

1. My thanks to Larry Arnhart, Patrick Bateson, Alan Bittles, and Mark Erickson for improving comments on my earlier draft.

2. Perry Anderson, *Considerations on Western Marxism* (London: New Left Books, 1976).

3. Claude Lévi-Strauss, *The Elementary Structures of Kinship*, trans. James Harle Bell and John Richard von Strurmer, and ed. Rodney Needham (Boston: Beacon Press, 1969), p. xxxix.

4. Marshall Sahlins, "Evolution: Specific and general," in *Evolution and Culture*, ed. Marshall Sahlins and Elman R. Service (Ann Arbor: University of Michigan Press, 1960), pp. 12, 38.

5. For example, David Schneider, *American Kinship: A Cultural Account*, 2nd ed. (Chicago: University of Chicago Press, 1980).

6. David Schneider, "The meaning of incest," *Journal of the Polynesian Society*, special issue on Incest Prohibitions in Micronesia and Polynesia (1976), pp. 149–70.

7. For example, Marvin Harris, *Our Kind: Who We Are, Where We Come From, Where We Are Going* (New York: Harper and Row, 1989). Cited in Paul R. Ehrlich, *Human Natures: Genes, Cultures, and the Human Prospect* (Washington, D. C., 2000), p. 201.

8. See Adam Kuper, *Culture: The Anthropologists' Account* (Cambridge: Harvard University Press, 1999).

9. Clifford Geertz, "Deep hanging out," *New York Review of Books*, October 22, 1998.

10. Bronislaw Malinowski, *Argonauts of the Western Pacific: An Account of Native Enterprise and Adventure in the Archipelagoes of Melanesian New Guinea* (London: Routledge and Kegan Paul, 1922); *Crime and Custom in Savage Society* (London: Kegan Paul, Trench, Trubner & Co., 1926); *Sex and Repression in Savage Society* (London: Kegan Paul, 1927); *The Sexual Life of Savages in North-Western Melanesia: An Ethnographic Account of Courtship, Marriage, and Family Life Among the Natives of the Trobriand Islands, British New Guinea* (London:

George Routledge & Sons, 1929); *Coral Gardens and Their Magic,* 2 vols. (London: George Allen & Unwin, 1935).

11. Bronislaw Malinowski, *Magic, Science, and Religion and Other Essays,* ed. Robert Redfield (Glencoe, Ill.: The Free Press, 1948); *A Diary in the Strict Sense of the Term* (New York: Harcourt, Brace, and World, 1967).

12. Malinowski, *Sex and Repression,* p. vii.

13. J. P. Singh Uberoi, *Politics of the Kula Ring: An Analysis of the Findings of Bronislaw Malinowski* (Manchester: University of Manchester Press, 1962); Marshall Sahlins, *Stone Age Economics* (Chicago: Aldine, 1972).

14. Annette B. Weiner, *Women of Value, Men of Renown: New Perspectives in Trobriand Exchange* (Austin: University of Texas Press, 1976); "Stability in banana leaves: Colonialism, economics, and Trobriand women," in *Women and Colonialization: Anthropological Perspectives* (New York: J. F. Bergin, 1980), pp. 270–93; *The Trobrianders of Papua New Guinea* (New York: Holt Rinehart, and Winston, 1988).

15. Weiner, *The Trobrianders,* p. 5.

16. For example, Malinowski, *Argonauts,* p. 399; *Crime and Custom,* pp. 116–17.

17. Weiner, *The Trobrianders,* pp. 25–31.

18. David Labby, *The Demystification of Yap* (Chicago: University of Chicago Press, 1976).

19. Karen Sacks, *Sisters and Wives: The Past and Future of Sexual Equality* (Urbana: University of Illinois Press, 1982).

20. Malinowski, *The Sexual Life of Savages,* pp. 425–26.

21. Ibid., p. 426.

22. Ibid., pp. 428–29.

23. Ibid.

24. Malinowski, *Argonauts.*

25. These are true yams (*Dioscorea spp.,* an ancient regional cultigen), not sweet potatoes (*Ipomoea spp.,* a post-Magellanic introduction). The continued use of yams in a major ritual of interdependence long after the adoption of many other staples suggests considerable age for this complex of sibling provisioning.

26. Weiner, *The Trobrianders of Papua New Guinea,* pp. 119–23.

27. Malinowski, *The Sexual Life of Savages,* p. 7.

28. Ibid., p. 440.

29. Ibid., p. 520.

30. Ibid., p. 519.

31. Even though Trobrianders passed on this shared myth, to assume that each individual "believed" literally in it is to attribute the primitive mass mentality among them that Malinowski was at pains to condemn and disprove.

32. Malinowski, *The Sexual Life of Savages,* p. 384.

33. Ibid., pp. 447, 450. These subclans are almost certainly those of the unusual Malasi clan, of whom more will be said below.

34. Ibid., p. 430.

35. Ibid., p. 433.

36. Ibid., p. 440.

37. Ibid., p. 441.

38. Ibid., pp. 384, 447.

39. Ibid., p. 446.

40. Ibid., pp. 74, 265, 445–48.

41. Ibid., p. 447.

42. Ibid., p. 390.

43. Ibid., p. 389.

44. Ibid., p. 424.

45. Ibid., p. 389.

46. Malinowski, *Crime and Custom*, p. 80.

47. Ibid., p. 81.

48. Malinowski, *The Sexual Life of Savages*, p. 439.

49. Malinowski, *Crime and Custom*, pp. 77–83.

50. Malinowski, *The Sexual Life of Savages*, p. 476.

51. Malinowski, *Crime and Custom*, pp. 118–19, 92.

52. Ibid., pp. 93–97.

53. Malinowski, *The Sexual Life of Savages*, p. 399.

54. Patrick Bateson, "Optimal outbreeding," in *Mate Choice*, ed. Patrick Bateson (Cambridge: Cambridge University Press, 1983). Also Chapter 1 of this volume.

55. Raymond Firth, *We, the Tikopia: A Sociological Study of Kinship in Primitive Polynesia* (London: Allen and Unwin, 1936), p. 340.

56. Weiner, *The Trobrianders*, pp. 8–9.

57. William Durham, *Coevolution: Genes, Culture, and Human Diversity* (Stanford, Calif.: Stanford University Press, 1991), p. 315.

58. Ibid., p. 333.

59. Ibid., p. 338.

60. Ibid.

61. Ibid., p. 357.

62. Durham is struck by the wide occurrence of the idea that incest results in defective children; this perception informs decision-making individuals, who then make culture of it. I find this a weak argument, not only for the traditional reasons (would small, sparsely distributed, sexually generous, and heavily inbred populations been able to interpret the genetic consequences of their matings?). Durham's interpretation may be a particularly good example of a problem that haunts thinkers from contracepting societies: we assume all too easily a distinction between recreational and reproductive heterosexual intercourse. Through most of human experience, children were an expected consequence of such intercourse, a part of the sexual experience for women at least. Having sex and making babies (even when the participants would rather not make a baby) were inextricably linked. Ethnography rapidly piles up examples of how various peoples contextualized intercourse (in the yam garden, often, after eating particular foods, according to ritual calendars) in order to ensure good-quality offspring. It is only a short conceptual step to predicting that "bad," incestuous sex would result in not-good children. I suggest that hypertrophied cultural logic rather than assessment of actual occurrence of inbreeding depression prompted the connection so often found in Durham's sample.

63. Durham, *Coevolution*, p. 359.

64. Mark Erickson, "Incest avoidance and familial bonding," *Journal of Anthropological Research*, vol. 45 (1989), pp. 280–95; "Rethinking Oedipus: An evolutionary perspective on incest avoidance," *American Journal of Psychiatry*, vol. 150 (1993), pp. 413–30; and Chapter 9 of this volume.

65. Jack Goody, "A comparative approach to incest and adultery," in *Comparative Studies in Kinship*, ed. Jack Goody (Stanford, Calif.: Stanford University Press, 1969), chapter 2.

66. Marvin Harris, *Cows, Pigs, Wars, and Witches: The Riddle of Culture* (New York: Random House, 1974), p. 5. Harris traveled a paradigmatic intellectual road from a truly vulgar materialism to a kind of Marxism that he labeled cultural materialism. To the end of his life, however, he sought to exclude human biology from this dialectic.

67. Morton H. Fried, *The Evolution of Political Society: An Essay in Political Anthropology* (New York: Random House, 1967); Jack Goody, *Production and Reproduction* (Cambridge: Cambridge University Press, 1976); Jack Goody, *The Oriental, the Ancient, and the Primitive: Systems of Family and Marriage in the Pre-Industrial Societies of Eurasia* (Cambridge: Cambridge University Press, 1990).

68. Stevan Harrell, *Human Families* (Boulder, Colorado: Westview Press, 1997), p. 27.

69. Ibid., pp. 129–30.

70. John M. Goggin and William C. Sturtevant, "The Calusa: A stratified, nonagricutlural society (with notes on sibling marriage)," in *Explorations in Cultural Anthropology*, ed. Ward H. Goodenough (New York: McGraw-Hill, 1964), pp. 179–220.

71. Ibid., pp. 204–6. Goggin and Sturtevant list eight "societies where (some type of) sibling marriage is not defined as incest for anyone" (p. 204). One of these at least—Hoklo—is easily eliminated; south Chinese minor marriages were made between adoptive, not biological, brothers and sisters. Of the others, the best known is that of Roman-period Egyptian commoners among whom full-brother-sister marriage was normative. Evidence for this society is strong but sparse. Its exceptionality is discussed by Wolf, who computes that, of 113 cases, only a dozen are demonstrably full siblings who can most reliably be assumed to have been reared together. The complexities of women's legal status as Hellenized Egypt came under Roman rule suggest that strong pressure for sibling marriage may have emerged in a conjuncture of status-climbing, ethnic, property-right, and marriage-law variants. See Keith Hopkins, "Brother-sister marriage in Roman Egypt," *Comparative Studies in Society and History*, vol. 22 (1980), pp. 303–54; Sarah B. Pomeroy, *Women in Hellenistic Egypt: From Alexander to Cleopatra* (Detroit: Wayne State University Press, 1990); and Chapter 5 of this volume.

72. Clifford Geertz, *Nagara: The Theater State in Nineteenth Century Bali* (Princeton, N.J.: Princeton University Press, 1980.

73. Cathy Small, "The political impact of marriage in a virtual Polynesian society," in *Dynamics in Human and Primate Societies: Agent-Based Modeling of Social and Spatial Processes*, ed. Timothy A. Kohler and George J. Gumerman (New York: Oxford University Press, 2000), p. 243.

74. Uberoi, *Politics of the Kula Ring*, pp. 31–36.

75. Malinowski, *The Sexual Life of Savages*, p. 420.

76. Ibid., pp. 385 and 447.

77. Ibid., p. 420.

78. William A. Lessa, *Ulithi: A Micronesian Design for Living* (New York: Holt, Rinehart, & Winston, 1966).

79. Malinowski, *The Sexual Life of Savages*, pp. 432, 456–74.

80. Ibid., pp. 474–75.

81. Arthur P. Wolf and Chieh-shan Huang, *Marriage and Adoption in China, 1845–1945* (Stanford, Calif.: Stanford University Press, 1980); Arthur P. Wolf, *Sexual Attraction and Childhood Association: A Chinese Brief for Edward Westermarck* (Stanford, Calif.: Stanford University Press, 1995); and Chapter 4 of this volume.

9 Evolutionary Thought and the Current Clinical Understanding of Incest

Mark T. Erickson

A distinct recollection of my psychiatric training in the 1980s is of a group of patients whose bewildering array of symptoms defied diagnosis. The records of these patients, who were mostly women, showed previous psychiatrists were equally puzzled. About this time surveys revealed the prevalence of incest to be far greater than imagined. This discovery generated a rapidly expanding research into the effects of incest.[1] Studies documented that incest, and frequently associated childhood neglect and physical abuse, predisposed individuals to a myriad of difficulties including posttraumatic stress disorder, chronic depression, alcohol and drug dependence, self-mutilation, suicide attempts, borderline personality disorder, anxiety disorders, somatoform disorders, bulimia nervosa, and dissociation.[2]

Patients who had been puzzling to clinicians, and to themselves, became more understandable. A fortunate consequence is that treatment of those suffering the effects of incest, and other trauma, has improved immeasurably.[3] Ironically, as clinicians found the prevalence of human incest to be greater than had been believed, biologists discovered that incest is rare in nature.[4] Further, anthropologists convincingly demonstrated that humans inherit an evolved propensity to *avoid* incest.[5]

Incest avoidance presumably evolved, in both human and nonhuman species, because of the harmful effects of close inbreeding.[6] Incest avoidance is not hardwired, or present at birth, but rather depends on close association between kin from early life. It is susceptible to disruption. Species that naturally avoid incest, including humans, are far more likely to engage in incest if early association is interfered with.[7]

The central purpose of this chapter is to reconcile the biology of incest avoidance with our current clinical understanding of incest. To begin, I briefly review literature on the biology of incest avoidance (more extensive reviews are found throughout this volume). Clinical research on incest will then be discussed within this broadened context. Examined in this manner, an integrative model of incest avoidance, as a biological adaptation, and incest, a pathological manifestation, emerges. The implications of this hypothesis are discussed. Many clinicians believe we are in the midst of an epidemic of incest. Although this assertion is difficult to prove, there is reason to believe that cultural practices, nonexistent in our evolutionary past, may disrupt our natural propensity for incest avoidance. This may increase incest prevalence—particularly the incestuous abuse of children.

Evolution and Incest Avoidance: A Brief View

For most of the twentieth century, it was widely accepted that a *cultural* rule, an incest taboo, was essential for inhibiting incest.[8] This view rested on the assumption that animals and precultural humans mated incestuously. Freud took this notion to its logical conclusion when he argued that humans are, by nature, incestuous. He proposed that repression of incestuous impulses created a universal neurosis, unique to our species. Freud called this neurosis the Oedipus complex.

Edward Westermarck, a Finnish anthropologist and contemporary of Freud, presented a strikingly different hypothesis.[9] Aware of the harmful effects of close inbreeding, Westermarck believed that through natural selection humans had acquired an *aversion* to incest. Westermarck, presciently, hypothesized that *close association from early life* established a later propensity for incest avoidance. Because children are raised in close proximity to parents, and siblings, in virtually all traditional cultures, his hypothesis was plausible.

Incest was not studied in nature until the 1960s. Given the long-held belief that it was common, primatologists studying rhesus monkeys were openly surprised when mother-son incest was found to be rare.[10] It has since been shown that incest is rare in other primate, mammalian, avian, amphibian, and even insect species.[11] With few exceptions, incest is uncommon throughout the animal kingdom.[12]

Consistent with Westermarck's hypothesis, early association between kin is essential to establishing incest avoidance in animal species.[13] Prairie voles, for example, rarely mate incestuously when reared naturally with siblings. If, however, siblings are separated at birth into foster litters, as adults they sexually avoid unrelated foster siblings but mate incestuously with unfamiliar biological siblings.[14]

The major obstacle to studying incest avoidance in humans is the incest taboo. Virtually all societies share this ostensibly culturally constructed rule, making it difficult to isolate biological influences.[15] To test Westermarck's hypothesis, it was necessary to find circumstances of early association without a later taboo on sexual affiliation. Two particularly useful test cases were found, one in Israel and the other in rural Taiwan.[16]

On Israeli communal farms, or kibbutzim, children of the same age were, until recent decades, raised together in a children's house from shortly after birth through high school graduation. Aside from visits with parents in the evening, children remained together both day and night. As they matured, no cultural rule opposed their sexual affiliation. If anything, peers were encouraged to date and marry. Because of early association, Westermarck's hypothesis predicts that peers should be sexually avoidant. This is what has been observed. Sexual relationships, whether through dating or marriage, are extremely uncommon between cosocialized peers.[17]

The most complete test of Westermarck's hypothesis is occurring in Taiwan, where Stanford anthropologist Arthur Wolf studies *simpua*, or "minor" marriage.[18] In minor marriage, the bride, or *simpua*, is usually betrothed in infancy and raised with her future husband in the groom's home. The couple is married in their midteens. Again, because of early association, Westermarck would predict that minor marriage couples will be averse to sexual affiliation, despite countervailing cultural pressures. Consistent with this prediction, Wolf has found that minor marriage couples have a much higher divorce rate, engage in adultery more frequently, and also have fewer children per year of marriage than couples in other arranged marriages. Wolf has meticulously ruled out alternative explanations of his data and concludes that Westermarck's hypothesis best comprehends the Taiwan findings.[19] Wolf's study indicates that the Westermarck effect, or incest avoidance, is dependent on close association during a sensitive period of about the first three years of life.[20]

As the twentieth century has come to a close, Westermarck's hypothesis has far more objective support than any alternative model. Incest avoidance studies provide a remarkably thorough test of this evolutionary hypothesis.[21] Nevertheless, clinicians are well aware that early association alone is not sufficient to establish incest avoidance. To the contrary, most incest occurs despite association. Given early association, are there definable influences that disrupt our natural propensity for incest avoidance? It is at this juncture where clinical research provides insights not found elsewhere.

Incest: Current Clinical Findings

PREVALENCE

Historically, estimates of the frequency of incest have varied to an extraordinary degree. In 1953, Alfred Kinsey et al. reported the prevalence of incest within the United States to be about one case per hundred.[22] In

TABLE 9.1
Estimated Prevalence of Incest, 1953–1996

Study	Year of Publication	Estimated Prevalence of Incest (percent)
A. C. Kinsey et al., *Sexual Behavior in the Human Female* (Philadelphia: W. B. Saunders).	1953	1.0 (all types)
S. K. Weinberg, *Incest Behavior* (New York: Citadel).	1955	0.0001 (all types)
D. Finkelhor, "Sex among siblings: A survey of prevalence, variety, and effects," *Archives of Sexual Behavior*, vol. 9, pp. 171–94.	1980	3.2 (sibling)[a]
D. E. H. Russell, "The incidence and prevalence of intrafamilial and extrafamilial sexual abuse of female children, *Child Abuse and Neglect*, vol. 7, pp. 133–46.	1983	2.9 (father-daughter) 2.2 (sibling)
G. Wyatt, "Sexual abuse of Afro-American and white-American women in childhood," *Child Abuse and Neglect*, vol. 9, pp. 507–19.	1985	1.6 (father-daughter)
A. W. Baker and S. P. Duncan, "Child sexual abuse: A study of prevalences in Great Britain," *Child Abuse and Neglect*, vol. 9, pp. 457–67.	1985	1.3 (all types)[a]
D. Finkelhor et al., "Sexual abuse in a national survey of adult men and women: Prevalence, characteristics, and risk factors," *Child Abuse and Neglect*, vol. 14, pp. 19–28.	1990	0.8 (all types)[a]
D. Finkelhor and J. Dziuba-Leatherman, "Children as victims of violence: A national survey," *Pediatrics*, vol. 94, pp. 413–20.	1994	0.9 (all types)[a]
H. Sariola and A. Uutela, "The prevalence and context of sexual abuse in Finland," *Child Abuse and Neglect*, vol. 20, pp. 843–50.	1996	0.2 (father-daughter)
D. S. Halperin et al., "Prevalence of child sexual abuse among adolescents in Geneva: Results of a cross sectional survey," *British Medical Journal*, vol. 312, pp. 1326–29.	1996	0.9 (all types)[b]

[a] In Finkelhor's 1980 study, 13 percent of the entire sample of 796 undergraduate students reported some type of sexual experience with a sibling (15 percent of females and 10 percent of males). Most of these experiences were regarded as childhood sex play. About 25 percent of this group, or a prevalence of about 3.2 percent, had experiences regarded as exploitative because force was used or because there was a large age difference between siblings. In this study and those by Baker and Duncan and by Finkelhor et al., the precise degree of relatedness was not specified. These data may include incest involving kin outside the immediate family or involving step-parents or step-siblings.

[b] All data cited from Halperin et al., "Prevalence of Child Sexual Abuse," are unpublished findings that were derived from the authors' 1996 study.

1955, Kirson Weinberg estimated the incidence at one case per million.[23] This estimate persisted and was cited as recently as 1975 in a widely used textbook of psychiatry.[24]

Since 1980 the prevalence of incest has been studied extensively. Table 9.1 summarizes the largest studies; combined, 17,045 individuals were surveyed (9,391 females and 7,654 males). As best possible, only incest within the nuclear family is included in Table 9.1. Step-fathers, uncles, cousins, and other nonimmediate kin were excluded. The reason for this exclusion was to find an approximation of the prevalence of incest, despite presumed early association, thereby providing a measure of how often the Westermarck effect, or incest avoidance, fails. It must be noted, however, that step-fathers, who are unlikely to be in close early association with step-daughters, are much more likely to perpetrate incest than biological fathers.[25] Further, the prevalence of early sexual abuse, from all possible sources, has been found to be far greater than imagined. Russell, for example, found that 28 percent of women, from a random sample, reported some form of sexual abuse before the age of fourteen.[26] Although child sexual abuse in general is not the focus of this chapter, this issue will be considered at the conclusion of the chapter.

The primary intent of prevalence surveys has been to determine the frequency of intrafamilial sexual encounters of any kind. The definition of incest therefore tends to be rather broad, often including everything from exhibitionism, to fondling, to intercourse. Of the studies listed in Table 9.1, five took place in the United States, one in Britain, one in Finland, and one in Switzerland. Prevalence studies are often very different in design. For example, some researchers surveyed adolescents at school with a pencil and paper questionnaire.[27] Others surveyed adults in a direct, face-to-face interview.[28] Telephone surveys were also used.[29] Despite methodological differences, the prevalence findings are in far closer agreement than earlier estimates.[30] Given the exceptionally sensitive nature of this subject it may be difficult to get more accurate findings.

Some surveys sought the prevalence of the most severe forms of incest. Daniel Halperin et al. found the prevalence of incest in which penetration occurred to be 0.3 percent.[31] Diana Russell defined the most severe form of incest as completed or attempted intercourse or oral sex.[32] Using this definition, she found a prevalence of 0.8 percent for both father-daughter and brother-sister incest. Anthony Baker and Sylvia Duncan found the prevalence of sexual intercourse "with a blood relative" to be 0.25 percent.[33] "Blood relative" was not clearly defined, so this finding may include persons beyond the immediate family. There are no studies that specifically examine the prevalence of mother-son incest in a general population. McCarty, however, found that 4 percent of a population of convicted sex offenders in a metropolitan treatment program were mothers who had sexually abused offspring.[34]

Several studies have examined the frequency of pregnancy caused by incest. Within the United States, estimated rates have varied from 1 percent to 20 percent. There appears to have been a decrease in incest pregnancy in recent decades as birth control has become more widely available.[35]

Incest prevalence data from non-Western cultures is virtually nonexistent. David Finkelhor has reviewed difficult-to-obtain data from twenty-one countries (i.e., unpublished or published in difficult-to-obtain sources). He found that across cultures females are most often the victims of incest, males are more typically perpetrators, and a history of sexual abuse is frequently associated with subsequent mental health impairment. No information on prevalence from non-Western cultures is given.[36]

INCEST WITH AND WITHOUT EARLY ASSOCIATION

To date, with few exceptions, clinicians have not used findings on the biology of incest avoidance to conceptualize clinical studies of incest.[37] The robust support for Westermarck's hypothesis indicates that the presence, or lack, of early association should be particularly informative. Questions not often considered by clinicians are raised. For example, incest taboos notwithstanding, does the lack of early association between kin make incest more likely? Is incest without early association phenomenologically distinct from incest with early association? Does incest with early association have recurrent characteristics? If so, do these characteristics suggest why the Westermarck effect is not invariable? Taken as a whole, clinical findings suggest answers to all of these questions.

Violations of the Westermarck Effect:
Incest with Early Association

Incestuous families often present a facade of respectability and may be overtly conventional to a fault.[38] With closer inspection, however, their apparent well-being is illusory. Clinicians have repeatedly found that intrafamilial behavior of incest families is pervasively disturbed.[39] In a controlled study, Philip Madonna, Susan Van Scoyk, and David P. Jones found that on a standardized family evaluation scale incest families tended to rate in the severely dysfunctional range. Among nonincest control families, there were clear boundaries between individuals, allowing for appropriate intimacy; these boundaries were far less distinct in incest families. Incest families were inefficient at resolving conflict. They were lacking in empathy. Parents tended to be neglectful, emotionally unavailable, and unable to support autonomy in offspring.[40] Alcoholism, drug abuse, and marital discord are more common in incest families.[41] Incest families tend to have a greater-than-average number of children.[42]

In father-daughter incest, fathers not infrequently offer their daughter gifts and special privileges to gain favor. This special relationship may be the daughter's only source of affection. When fathers pursue incest, however, the daughter's experience is almost invariably one of fear, disgust, disbelief, confusion, anger, and shame.[43] In her study of father-daughter incest, Patricia Phelan found that none of the daughters initiated the activity or enjoyed what happened.[44] Given early association, it is only on rare occasions that incest appears to be emotionally acceptable to a daughter.[45]

Mothers, consciously or unconsciously, are often complicit with father-daughter incest. In subtle ways a mother may encourage, or at least not discourage, her husband's incestuous behavior. Daughters who confide in their mothers are usually bitterly disappointed. Mothers, even when fully aware of paternal incest, often do not defend their daughters. Not surprisingly, victims of father-daughter incest have overwhelmingly negative images of their mothers. In Judith Herman's study, thirty-nine of forty daughters who were victims of father-daughter incest had extremely negative images of their mothers, describing them as cold, indifferent, and ungiving. The only exception was a daughter who had lost her mother in early childhood.[46]

Because mother-child incest is so uncommon, information had been extremely limited, consisting of little more than case reports until the studies of Loretta McCarty and Kathleen Faller.[47] Combined, these authors gathered data on sixty mothers convicted of sexual abuse of offspring. Both found patterns of extreme pathology within the family. Neglect and physical abuse often accompanied maternal incest.

Some writers have argued that social isolation and poverty may be critical factors underlying incest.[48] Other studies contradict this notion.[49] It has been proposed that male dominance in a paternalistic society lies at the root of incestuous behavior.[50] More recent studies do not support this view.[51]

The most salient influence on incest behavior may be found in the childhood attachment experience of parents of incest families.[52] Incestuous fathers, for example, typically describe their childhood as filled with rejection, neglect, and physical and/or sexual abuse. Parental absence because of death or abandonment is also common.[53] In father-daughter incest families, the mother's early experience is similarly bleak. She is likely to have had an emotionally deprived childhood characterized by rejection and hostility. A history of childhood sexual abuse is frequently found.[54]

McCarty presents the most extensive information on the childhood of incestuous mothers. In her study nearly all mothers described an unremittingly bleak childhood using adjectives such as "rough" or "horrible" to describe their upbringing. Physical and/or sexual abuse were extremely common (95 percent) during the childhood of these mothers.[55]

Similar developmental conditions are observed in sibling incest. The childhood of the offender, most frequently an older brother, and the victim, usually a younger sister, is typically overwhelmingly grim. Mothers are described as "emotionally absent," "distant," "inaccessible," or "neglectful."[56] Likewise, fathers are often absent, through death or by abandonment or divorce. If present, the father is usually emotionally distant.[57] Naomi Adler and Joseph Schultz found that 92 percent of boys who perpetrated incest had been physically abused by one or both parents.[58] Frequently, such boys have also been sexually abused.[59] The personal relationship between an offending brother and his sister has been described as nonexistent except for incest and physical abuse.[60]

A particularly appalling finding is that incest is often initiated very early in the victim's life. Father-daughter incest may begin when the daughter is six years of age or younger.[61] The average age of onset in father-daughter incest is about eight to nine years.[62] In a study of mother-child incest, the mean age of victims *at assessment was* 6.4 years.[63] In sibling incest the brother typically initiates sexual abuse when he is between eleven and fourteen years of age. The mean age of the sister at onset is about seven years.[64] The early onset of much of human incest does not appear to be linked to fixated pedophilia. Perpetrators of incest rarely limit their sexual attention to children.[65]

Incest inflicted on children, and more broadly, child sexual abuse, appears to be a uniquely human behavior. This has *not* been observed in other primate species.[66] Given that the conditions for establishing incest avoidance, association during a sensitive period, appears to be similar across mammalian species, this variation is cause for concern. Incest perpetrated on children may reflect a biological peculiarity of our species. A more plausible explanation is that this propensity is due to cultural influences, rare or nonexistent in our evolutionary past, which interfere with the development of incest avoidance. The incest avoidance adaptation can be easily disrupted in animal species by artificial intrusion.[67] There is little reason to believe we are an exception.

Incest Following Early Separation: A Phenomenologically Distinct Entity

In his large study, Weinberg expressed particular interest in six pairs of incestuous siblings.[68] Each sibling of all six pairs desired the incestuous relationship. There was no evidence of coercion on the part of the brother as is usually the case in sibling incest. Though aware of an incest taboo, these siblings were largely indifferent to this injunction. They lacked apparent feelings of guilt concerning their relationship. Relationships often began quickly and were passionate. Three of the couples eventually married. Remarkably, every pair had been separated from early childhood and were only later reunited.

In 1975 Britain enacted the Access to Birth Records Act. This act enables adopted individuals over the age of eighteen to trace biological kin. An unanticipated complication of the Birth Records Act has been the erotic feelings frequently experienced by reunited kin. A study by Maurice Greenberg and Roland Littlewood suggests that incest may frequently occur.[69] They estimate that over 50 percent of reunited kin experience strong sexual feelings. The following examples from their study are illustrative:

(A 22 year old civil servant meets his mother) "We meet, smile, kiss. . . . I notice her nose, brow, deep set eyes; I am constantly looking for similarities. . . . She said 'I've got to touch you' and touched my face with her hand. It felt nice. . . . Monday we dressed up for each other. . . . She looked gorgeous, I felt teased, and she admitted she was teasing me. . . . We kissed tinkering on the edge."

(A 35 year old nurse meets her biological father) "It developed very quickly, we hugged and kissed a lot that first weekend. His skin felt like mine and he smelled like me. I had a sexual dream about him, wanting it. I just thought it was crazy but discovered he was open to it."

Many similar anecdotal reports exist. "Susan," who was adopted at birth, sought her biological parents at age twenty-two. Within months she located her father. When asked about her reunion she commented, "There was an immediate sense of recognition. He looked very like me . . . the face, the gestures. . . . I knew I felt attracted to him which really scared me. I never talked to him about it but I believe he felt the same."[70]

Allen and Patty Muth, brother and sister, had an incestuous relationship that spanned several years and resulted in the birth of four children. They moved from one region of the United States to another, staying ahead of criminal charges. The couple was eventually apprehended in Wisconsin, convicted, and sentenced for the felony of incest. Mentioned virtually as an aside in the popular press, was that the couple did not meet until Patty, the younger sibling, was eighteen years of age.[71]

Victoria Pittorino found her brother, David Goddu, through birth records of the State of Massachusetts. They were separated by adoption when Victoria was three and David was one year old. Twenty years later they reunited. Ms. Pittorino described her response as "love at first sight." The attraction was mutual, and the couple married within weeks of reunion. Later, they were arrested and convicted of incest under Massachusetts statutes that originated in 1695.[72]

"Jackie," an adoptee, found her biological brother. She related, "I felt irresistibly, magnetically drawn. . . . My body reaction was to move toward him. I wanted eye contact. I liked what I saw because he was like me."[73]

At the age of twenty-five, the poet Byron entered what has been described as a torrid affair with his half-sister Augusta Leigh. Those who saw them together were struck by their resemblance. They met for the first time in 1801, when Byron was fourteen and Augusta seventeen.[74]

Perhaps the most vivid description of father-daughter incest is found in the diaries of writer Anaïs Nin. Although not completely separated from her father, concert pianist Joaquin Nin, their early contact was significantly limited because Joaquin was frequently on tour. When at home, he withdrew from the life of his family. When Joaquin did notice Anaïs at all, it was usually to call her the "ugly little girl."[75] When Anaïs was ten years old, Joaquin Nin abandoned his family altogether. Twenty years later, Anaïs and her father reunited. "But you are like me," said Joaquin, who then commented on the similarities of their hands and hair.[76] They meet again, at which time Joaquin states, "You are the synthesis of all women I have loved. . . . I don't feel towards you as if you were my daughter." Anaïs replied, "I don't feel as if you were my father." He responded, "I have finally met the woman of my life and it is my daughter. . . . I'm in love with my own daughter." Anaïs replied, "Everything you feel is reciprocal."[77]

A recurrent story line in poems, novels, and plays in which incest is a theme is one of early separation.[78] Examples include *The Book Bag* by Somerset Maugham, *Pierre, or the Ambiguities* by Herman Melville, and *The Caryatids* by Isak Dinesen. The myth of Oedipus is the best-known example. Oedipus is separated from his mother, Jocasta, at birth. Many years later they unknowingly reunite incestuously. Clinical and anthropological findings now illuminate this myth in a way not anticipated by Freud. Oedipus portrays the literal truth that early separation undermines a natural adaptation for incest avoidance.

As predicted by Westermarck's hypothesis, clinical studies show that incest is far more likely if kin are separated early in life. It is, virtually, only in this circumstance that incest may be mutually desired and eventuate in marriage.[79] Taboos appear to have limited influence. By contrast, given early association, incest is rarely, if ever, mutually desired. It is perpetrated coercively, by fathers or brothers, and experienced as being intensely aversive by daughters or sisters.

One clinical finding not predicted by Westermarck is the extraordinary fascination reunited kin show for each other. They often describe an immediate sense of recognition. They notice that they smell alike, that their gestures are similar, that they "resemble each other in ways that transcend physical similarities."[80] The basis of this unusually intense attraction is far from clear, but it may derive from a tendency, observed in many species, to respond preferentially to similarities between self and others. Biologists call this "phenotypic matching."[81]

Many species, including humans, quickly learn to identify familial characteristics; at birth, mothers typically adopt an *en face* orientation for close inspection of their infant. Within a very short time, they can reliably identify their offspring through visual, auditory, olfactory, or tactile cues.[82] Similarly, fathers appear to have a remarkable ability to identify offspring soon after birth. With an average of less than seven hours of postnatal contact, blindfolded fathers were able to recognize offspring by touch of the infant's hand alone.[83] Infants also quickly develop a preference for the parent's phenotype. Two-week-old breast-feeding infants, for example, are capable of recognizing their mother by olfactory cues alone.[84] Under normal developmental circumstances, early association generates a later sexual preference for someone who is rather different from close kin.[85] Separated kin, in contrast, may be extraordinarily intrigued by their similarities because of an inherent tendency for phenotypic comparison or matching.[86] Because of their lack of early association, however, a sense of kinship and incest avoidance are not established. Joaquin and Anaïs Nin, for example, recognized that they did not "feel" as though they were father and daughter, yet expressed intense interest in their physical similarities and engaged in incest.

A second clinical finding not anticipated by Westermarck's hypothesis is the very high frequency of abuse and neglect during the childhood of the parents of incest families. This suggests that the *propensity* for later incest may be influenced very early in life by the quality of attachment relationships. This notion will be developed in the next section.

Incest and the Psychopathologies of Kinship

All biological adaptations are susceptible to pathology. Bones break, the blood supply to the heart becomes occluded, immune systems attack the bodies they were designed to protect. Incest avoidance, a biological adaptation, is no exception.

Understanding normal development of any adaptation invariably leads to a better understanding of pathology. Westermarck's hypothesis says little about the development of incest avoidance aside from the necessity of early association. Can more be said? Incest avoidance might, for example, develop independently with no connection to other classes of social behavior. If this were the case, however, incest should be observed as an isolated event in otherwise healthy families. Clinical studies are clearly contradictory. Incest occurs, overwhelmingly, in grossly disturbed families in which neglect, abandonment, and physical abuse are also common. Incest may then reflect pathology in an adaptation that broadly modulates adaptive behaviors of kinship.

EVOLUTION AND KINSHIP

For Freud and many of his contemporaries, kinship was a uniquely human phenomenon resting precariously on a cultural taboo that restrained incestuous and aggressive instincts. A revolutionary change in our understanding of kinship and its evolutionary underpinnings began in the 1960s with the work of biologists W. D. Hamilton, John Maynard Smith, and G. C. Williams.[87] A useful, if somewhat oversimplified, summation of their argument is that natural selection maximizes the ability of individual organisms, not species, to gain *genetic representation* in future generations.[88] Because individuals are genetically more similar to kin than nonkin, this notion predicts that individual organisms will maximize their genetic representation by showing preference to kin.

Altruistic behavior is the most obvious class of behavior predicted to occur between close kin. Parents who, for example, show a preference for their own offspring in providing food, shelter, and so forth should more effectively maximize their genetic representation than parents who show no such preference. Attachment behaviors of the young are predicted to be kin-directed because kin should be most inclined to respond altruistically to offspring. Sexual (incest) avoidance is a third class of behavior that should be kin-directed.

Behavioral studies in recent decades have revealed that the phylogenetic roots of kinship go far deeper than believed by Freud. From the kin-directed warning calls of Belding ground squirrels, to exclusive maternal preference for offspring in California sea lions, to nepotism among rhesus monkeys, studies have repeatedly shown that kinship powerfully predicts patterns of social behavior.[89] Altruistic behaviors have been found to occur predominantly between immediate kin.[90] British psychiatrist John Bowlby found that in humans attachment bonds are, across cultures, directed primarily to biological parents.[91] Incest avoidance, as previously discussed, has been documented in a vast range of species including humans. The discovery that social organisms, from insects to primates, respond preferentially and nonsexually to kin raises questions which are ultimately of great clinical importance. If kinship is not a cultural construct, but rather has an evolutionary history that long precedes societal taboos and our species, then how do kin recognize one another? How do the distinctive patterns of kinship behavior develop?

Patrick Bateson of Cambridge University initially proposed that sensitive periods, long described by ethologists, could play an essential role in the development of kin-recognition.[92] Indeed, association during sensitive periods now appears to be the most common mechanism by which kin come to recognize one another.[93] This process is overtly simple. Association during an early sensitive period canalizes stable social bonds from child to parent, parent to child, and sibling to sibling. Over a century ago Westermarck proposed that early association was the foundation of human incest avoidance. It now seems that early association functions far more broadly, canalizing different classes of kin-directed behavior, including attachment, altruistic behavior, and incest avoidance, across many species.

INCEST AVOIDANCE AND KINSHIP

Taking development one step further raises the question of whether different classes of kin-directed behaviors function independently or, conversely, are integrated into an overarching system. For reasons both theoretical and empirical, Arthur Wolf and I have hypothesized that kin-directed behaviors are either integrated or develop so closely that, for all intents, they are functionally integrated.[94] This proposal significantly expands upon Westermarck's hypothesis and leads to new, testable predictions about the relationship between incest avoidance, attachment, and kin-directed altruism.

Theoretically, if evolutionary forces increase the social interdependence of kin, thereby prolonging affiliation, the adaptive significance of incest avoidance increases accordingly. The *coevolution* of attachment, kin-directed altruism, and incest avoidance therefore seems probable.[95] To give two suggestive examples, male pilot whales swim in their mother's group, or pod, throughout their adult life. There they remain in close proximity to their mother and other kin, including mature sisters. Males leave their pod only transiently, to mate. Lifelong association notwithstanding, genetic studies of entire pods show that incest is rare, if not nonexistent.[96] Male bonobo chimpanzees develop particularly strong attachments to their mother and maintain close ties throughout their lives. Mature mother-son pairs migrate together and are mutually supportive. Despite ongoing close association, and the active sexual lives of bonobos, incest between mother-son pairs is rare.[97]

A second theoretical point comes from the work of French geneticist Francois Jacob.[98] Jacob, in a classic lecture given at Berkeley, observed that "evolution does not produce novelties from scratch" but instead utilizes what already exists, transforming this into more elaborate systems. According to Jacob, it would be unlikely that separate mechanisms of kin recognition evolved for each class of familial behavior. It is more probable that once an adaptation for kin recognition evolved (association), this was utilized by natural selection to canalize different classes of kin-directed behavior, including attachment, kin-directed altruism, and incest avoidance.

A third reason to suspect that kin-directed behaviors are integrated comes from attachment research. Secure attachment develops when parents are *responsive to* an infant's needs.[99] Conversely, insecure attachment emerges when parents are unresponsive. Individuals who develop secure attachments in their own childhood (as judged by the Adult Attachment Inventory) tend as adults to provide responsive parenting and have securely attached offspring (as measured by the Ainsworth Strange Situation). Conversely, parents who had insecure attachments in their childhood are usually less responsive to offspring, who in turn become insecurely attached.[100] It is important to recall that the early childhood experience of incestuous fathers, mothers, and siblings is marked by neglect, abandonment, and physical and sexual abuse. These are the conditions that lead to highly insecure attachment. Thus, the very conditions that contribute to insecure attachment in childhood appear to be linked to a later propensity for unresponsive parenting and incestuous abuse.

THE PSYCHOLOGY OF INCEST AVOIDANCE

Westermarck did not propose a psychology of incest avoidance. Freud, of course, did, and his Oedipal psychology was based on the assumption that all social bonds are ultimately sexual. This assumption created the dilemma of explaining how sexuality is inhibited among kin. Freud argued that an aversive experience, childhood castration fear, was essential. Decades later, with the discovery of a biological basis for incest avoidance, a renewed interest in the psychology of incest avoidance emerged. Initially, authors continued within the Freudian paradigm.

Anthropologist Robin Fox, for example, proposed that childhood sexual play between siblings is frustrating because of an inability of children to reach orgasm.[101] He argued that this frustrating sexual play would act as a negative reinforcement to later incest among siblings. Fox's hypothesis has several limitations, including a lack of evidence that childhood sexual play is frustrating and evidence that children can indeed achieve orgasm.[102] In a similar vein, anthropologist William Demarest proposed that punishment in the childhood home acts to aversively condition incestuous behavior.[103] Although severe early punishment can inhibit later sexual behavior in animals, the effect is general, not specific, as would be needed for adaptive incest avoidance.[104] Beyond these objections are clinical findings showing that incest is *most* prevalent in families where aversive experiences are particularly common—the opposite of what this paradigm would seem to predict.

Psychodynamically oriented authors in disagreement with the Oedipal model have argued, or inferred, that under normal developmental conditions a nonsexual form of affiliation exists between kin.[105] These authors, however, lacked a coherent explanation of how nonsexual affiliation could emerge, and formal hypotheses were not put forth.

A crucial shift in conceptualizing the psychology of incest avoidance begins with the recognition that a distinct "familial" affiliation can evolve because selectional forces shaping kin-directed behaviors (altruism, attachment, incest avoidance) are quite different from those shaping sexual or pair bonding.[106] Distinguishing familial and sexual affiliation resolves the difficulty of explaining incest avoidance as a product of aversive experience and psychological repression. Rather, incest avoidance begins, psychologically, with experiences that establish familial affiliation. Attachment studies suggest this experience is a responsive childhood milieu, which, in turn, elicits secure attachment.

The kibbutz peer findings are illustrative. Peers develop early mutual attachments that appear to be quite secure.[107] These attachments remain stable from childhood into adulthood, as evidenced by the intimate, responsive, and nonsexual relationships that endure between peers.[108] This "familial bond" between kibbutz peers develops at an unconscious level, as it does in other species. When kibbutz peers are asked why they are not mutually sexually attracted, they commonly reply that they experience each other as brother and sister. Words such as brother, sister, filial, paternal, and maternal connote feelings usually bestowed on family members. In using such terms, peers reveal how language points to the existence of a familial type of social affiliation that is biologically and psychologically distinct from sexual affiliation.

Conversely, early abuse and neglect undermine the propensity for adaptive familial affiliation at an equally unconscious level. Allen Sroufe and Mary Ward videotaped the interactions of 173 mothers with their toddlers. Most mothers interacted with warmth, empathy, and affection, but about 9 percent of the mothers were observed to be "seductive" or sexually inappropriate with their child, *even while being videotaped*. A mother might, for example, ask her child for a kiss and then, if refused, physically force a "passionate" kiss on the child's lips. Seductive mothers were also more likely to physically abuse or strike their child. A subset of all mothers were interviewed about their own childhood. Mothers whose childhood history was suggestive of incest were more likely to behave in a seductive and physically abusive way with their children. The authors note that while "affectionate" mothers were responsive to offspring, "seductive" mothers were noticeably less responsive and often interrupted the flow of the child's behavior.[109] Clare Haynes-Seman and Richard Krugman report similar observations. Parents who had reported a childhood history of neglect or abuse tended to exhibit sexualized interactions with their infants even while aware the session was being videotaped. These studies suggest that confusion about appropriate parental care and sexual boundaries is, to an extraordinary degree, unconscious.[110] The basis of this boundary confusion appears to lie in insecure childhood attachment.

To give one last example, Herman describes a group of fathers who, while not overtly incestuous, sexualized the relationship with their daughters.[111] These fathers might present their daughters with "sexy underwear," frequently talk about sex with their daughters and so forth. Emotionally, such families tended to be tense and cold. Physical displays of affection were uncommon and uncomfortable. The relationship between daughter and father existed not in secure affection but in a milieu of distrust. The attachment hypothesis of incest avoidance would predict that seductive fathers did not develop an adequately secure attachment during their own childhood. They experience unconscious boundary confusion between sexual and familial affiliation but less so than overtly incestuous fathers. This prediction could be evaluated with instruments such as the extensively used Adult Attachment Inventory.[112]

Conceptualized in this way, the extraordinary psychological virulence of incest becomes more understandable. Familial and sexual affiliation evolved into biologically distinct entities because each is adaptive within a different social context. Disrupting normal development, through incest, physical abuse, or neglect, undermines adaptive social functioning at its most basic level. Later in life, secure, intimate relationships, whether with one's children or a spouse, become extraordinarily difficult to achieve. It is small wonder that borderline personality disorder, major depression, anxiety disorders, drug and alcohol dependence, and chronic suicidality are frequently observed sequelae.

A discussion of the psychology of incest avoidance would not be complete without mentioning "infantile sexuality." Freud placed great emphasis on this subject. Yet even within psychoanalytic circles the scientific validity of Freud's notions has been openly questioned.[113] Early sexual play is observed in other primate species. Infant and juvenile chimpanzees, for example, will mount their mother.[114] The significance of this behavior is unknown—it might facilitate the development of adult sexual behavior or have a soothing function.[115] There is remarkably little controlled research on early sexual play in humans. Although most childhood sexual play is probably within a developmental norm for our species, attachment studies suggest that the quality of early attachment directly influences early sexuality. Sroufe et al., for example, found that insecure attachment predisposed to precocious, sexualized behavior and boundary violations in preadolescence. Secure attachment was not associated with such behavior.[116]

One final point must be made concerning the boundaries between familial and sexual affiliation. Natural selection opposes not only close inbreeding but also excessive outbreeding.[117] In light of this, one might expect to find individuals preferring mates who are neither too similar nor too different from themselves. Consistent with this, Bateson has shown that Japanese quail are most sexually attracted to birds whose feather coloration is somewhat, but not dramatically, different from that of immediate kin.[118] Quail "sexually imprint" on the coloration of immediate kin and use this, comparatively, for mate selection. Claus Wedekind et al. provide evidence suggestive of a similar process in humans. In this case, body odor, as assessed by differences in the major histocompatibility complex (MHC), influenced sexual preference. Females found the odor of a male more attractive if his MHC, and hence body odor, was relatively different from her own.[119] Recent findings from cross-cultural research and endocrinology suggest the propensity for sexual affiliation develops much later than that for familial bonding.[120]

To summarize, human incest avoidance appears to depend on at least two factors: (1) early association between kin, and (2) adequately secure childhood attachment. These two factors establish the propensity for adaptive familial affiliation. The powerful influence of association during a sensitive period is documented by Wolf's *simpua* marriage research in Taiwan, the Israeli kibbutz peer studies, and clinical findings on incest following early separation.[121] It is further supported by animal studies showing that incest avoidance in other species depends critically on early association.[122]

The importance of adequately nurturant parenting is suggested by the developmental histories of incest families. Clinical research has repeatedly found that the parents of incest families were, themselves, neglected, abused, or abandoned as children.[123] This is not to say that an abusive, neglectful childhood invariably leads to later incestuous behavior. Many, perhaps most, individuals survive harsh childhoods and go on to be good parents. Rather, it appears that early abuse and neglect are potent factors for disrupting early attachment and later propensities for parental care and incest avoidance.

DO MODERN CULTURAL PRACTICES MAKE INCEST MORE COMMON?

Many clinicians question whether we are in the midst of an epidemic of incest. If even marginally true, this is cause for concern given the pervasively harmful effects of incest. Prevalence studies provide no answer. All research is too recent to offer historical measure. There is, however, a different approach to this concern. As mentioned, a unique and particularly troubling characteristic of human incest is that it is frequently perpetrated on children. This behavior is not observed in other primate species.[124] Incestuous abuse of children, and child sexual abuse in general, may be a product of modern cultural practices. These cultural practices, novel to our evolutionary past, may severely disrupt early attachment and, consequently, the development of unconscious, adaptive boundaries between familial and sexual affiliation.

The human genome has changed little since our species emerged 100,000 to 150,000 years ago. We remain nearly identical to our Paleolithic ancestors yet live in a dramatically different world. This mismatch between our evolved biology and the modern world has, almost certainly, altered patterns of human pathology. The current diet of most Westerners, for example, includes a much higher intake of salt than in hunter-gatherer cultures. Cultures whose diet is free of added salt have low blood pressure, by American standards, and do not show an increase in blood pressure with age as is common in industrialized society. Hypertension is rare.[125] Analogously, cultural practices in Western industrialized societies may interfere with early attachment to such an extent that behaviors such as child-directed incest manifest.

Until the twentieth century virtually all babies were delivered in the home. By 1970, nearly all births in the United States occurred in hospitals. Many clinicians and mothers have questioned whether modern birthing practices interfere with the mother-infant relationship at its earliest stages. Studies now show that simply providing an undivided source of emotional support for a mother during labor and delivery can dramatically reduce the Cesarean section rate. Women who receive emotional support tend to spend more time stroking, smiling at, and talking to their babies immediately following delivery than women who receive routine hospital care. Weeks later, supported mothers are observed to spend more time with and have more positive feelings for their infant.[126] In a randomized study, Susan O'Connor et al. found that mothers who "roomed in" with their infant, in the hospital during the first two days postpartum were at follow-up, twelve to twenty-one months later, significantly less likely to have abused or neglected their infant than mothers who received routine postpartum care.[127] Although the studies of O'Connor and others have been controversial, recent research reveals similar findings in other primate species.[128] A review of the literature by Dario Maestripieri suggests the existence of a postpartum sensitive period of heightened "maternal motivation," which, in effect, canalizes a maternal bond to offspring.[129] Separation from offspring during this period increased the probability of maternal rejection among other primates. This effect may be mediated, in humans and other primates, by hormones such as oxytocin and prolactin.[130]

Mammalian breast-feeding has been molded by more than 65 million years of evolution. In addition to its nutritional function, breast-feeding may also affect emotional development. Primate cross-species comparisons suggest that human children have evolved to receive benefits from breast-feeding for an absolute minimum of two and a half years to an upper limit of seven years.[131] In the United States, many women do not nurse at all, and physicians often consider six months to be extended breast-feeding. Modern breast-feeding practices undoubtedly reduce physical if not emotional intimacy between mother and child.

In Western industrialized countries, infant crying tends to be prolonged. The duration of crying bouts is shorter in cultures such as the !Kung San of Botswana, where infants are carried virtually continuously in a sling. !Kung San mothers, by experimental measures, have been rated as much more responsive than Western mothers. Infants are nursed on demand rather than by "schedule." This indulgent form of caregiving by the !Kung San is probably typical for our species and other higher primates. Frequent short cries elicit a positive, communicative response. Repeated many times a day, these may engender in the infant a secure, confident attachment with mother.[132]

In our evolutionary past, parents slept in immediate proximity to infants and young offspring. This practice not only provided protection but also appears to promote normal physiological development in infants and may reduce pathology such as sudden infant death syndrome.[133] In the vast majority of non-Western cultures, various forms of parent-child cosleeping are the predominant arrangement throughout the first few years of life. The practice of placing infants and young children into separate bedrooms is an entirely anomalous cultural intrusion. This practice obviously decreases the amount of close association between parent and child. Apropos of this, John Forbes and David Weiss found that, "contrary to expectations," children who coslept with parents were *less* likely to have been treated in a mental health clinic for emotional and behavioral problems.[134]

Diminished involvement of the father, or stepfather, in the caretaking of children is associated with increased risk of paternal incest. Hilda Parker and Seymour Parker found that sexually abusive fathers were less likely to have been involved in the early care of their daughters.[135] Linda Williams and David Finkelhor similarly found that low involvement by fathers in caretaking of offspring was a risk factor for incest. They note that the act of taking care of a child may evoke feelings of nurturance that are incompatible with incestuous abuse.[136]

Early association is crucial for the development of adaptive familial bonds in many species. Humans are no exception. The examples just mentioned represent only a few of the ways in which the quality and quantity of early kin association may be disrupted by cultural practices that have no precedent in our evolutionary past. The combined effects of such practices may manifest as severe pathologies of kinship. Incest perpetrated on children may represent the most extreme example.

During most of the twentieth century, social scientists believed incest was common in nature. Among humans, incest was thought to be rare because of cultural taboos. As the twenty-first century begins, this view has been, figuratively, turned on its head. Incest is now known to be rare in nature, and we must ask if human incest has become more, not less common *because* of cultural influences. If pathologies of kinship exist, antidotes will be found. As with medical science in general, the speed of discovery will depend on how findings from nonclinical disciplines are used to inform clinical research. Efficient progress requires open communication between the varied disciplines of anthropology, psychiatry, psychology, ethology, behavioral endocrinology, and all other fields that contribute to our understanding of human kinship and familial relations.

NOTES

1. Erna Olafson, David L. Corwin, and Roland C. Summit, "Modern history of child sexual abuse awareness: Cycles of discovery and suppression," *Child Abuse and Neglect*, vol. 17 (1993), pp. 7–24.

2. Arthur J. Barsky, Carol Wood, Maria C. Barnett, and Paul D. Clay, "Histories of childhood trauma in adult hypochondriacal patients," *American Journal of Psychiatry*, vol. 151 (1994), pp. 397–401; James A. Chu and Diana L. Dill, "Dissociative symptoms in relation to childhood physical and sexual abuse," *American Journal of Psychiatry*, vol. 147 (1990), pp. 887–92; Kenneth S. Kendler, Cynthia M. Bulik, Judy Silberg, John M. Hettema, and John Myers, "Childhood sexual abuse and adult psychiatric and substance use disorders in women," *Archives of General Psychiatry*, vol. 57 (2000), pp. 953–59; Pamela S. Ludolph, Drew Westen, Barbara Misle, Anne Jackson, Jean Wixom, and F. Charles Wiss, "Borderline diagnosis in adolescents: Symptoms and developmental history," *American Journal of Psychiatry*, vol. 147 (1990), pp. 470–76; James Morrison, "Childhood sexual histories of women with somatization disorder," *American Journal of Psychiatry*, vol. 146 (1989), pp. 239–41; Susan M. Ogata, Kenneth R. Silk, Sonya Goodrick, Naomi E. Lohr, Drew Westen, and Elizabeth M. Hill, "Childhood sexual and physical abuse in adult patients with borderline personality disorders," *American Journal of Psychiatry*, vol. 147 (1990), pp. 1008–13; Elizabeth F. Pribor and Stephen H. Dinwiddie, "Psychiatric correlates of incest in childhood," *American Journal of Psychiatry*, vol. 149 (1992), pp. 52–56; Elizabeth F. Pribor, Sean H. Yutzy, J. Todd Dean, and Richard D. Wetzel, "Briquet's syndrome, dissociation, and abuse," *American Journal of Psychiatry*, vol. 150 (1993), pp. 1507–11; Donna H. Schetky, "A review of the literature on the long-term effects of childhood sexual abuse," in *Incest-Related Syndromes of Adult Psychopathology*, ed. Richard P. Kluft (Washington, D.C.: American Psychiatric Press, 1990), pp. 35–54; Steven L. Shearer, Charles P. Peters, Miles S. Quaytman, and Richard L. Ogden, "Frequency and correlates of childhood sexual and physical abuse histories in adult female borderline patients," *American Journal of Psychiatry*,

vol. 147 (1990), pp. 214–16; Michael H. Stone, "Incest in the borderline patient," in *Incest Related Syndromes of Adult Psychopathology*, ed. Richard P. Kluft (Washington, D.C.: American Psychiatric Press, 1990), pp. 183–204; Bessel A. van der Kolk, Christopher Perry, and Judith L. Herman, "Childhood origins of self-destructive behavior," *American Journal of Psychiatry*, vol. 148 (1991), pp. 1665–71; Edward D. Walker, Wayne J. Katon, Kathleen Neraas, Ron P. Jemelka, and Donna Massoth, "Dissociation in women with chronic pelvic pain," *American Journal of Psychiatry*, vol. 149 (1992), pp. 534–37; Edward D. Walker, Wayne J. Katon, Peter P. Roy-Byrne, Ron J. Jemetka, Jean Russo, et al., "Histories of sexual victimization in patients with irritable bowel syndrome," *American Journal of Psychiatry*, vol. 150 (1993), pp. 1502–6.

3. See Jon Allen, *Coping with Trauma: A Guide to Self Understanding* (Washington, D.C.: American Psychiatric Association Press, 1995); Jon Allen, *Traumatic Relationships and Serious Mental Disorders* (New York: Wiley, 2001); Judith L. Herman, *Trauma and Recovery* (New York: Basic Books, 1992).

4. See Anne E. Pusey and Marisa Wolf, "Inbreeding avoidance in animals," *Trends in Evolutionary Ecology*, vol. 11 (1996), pp. 201–6; and also Chapter 3 of this volume.

5. Joseph Shepher, "Mate selection among second-generation kibbutz adolescents and adults: Incest avoidance and negative imprinting," *Archives of Sexual Behavior*, vol. 1 (1971), pp. 293–307; Arthur P. Wolf, "Childhood association, sexual attraction, and the incest taboo: A Chinese case," *American Anthropologist*, vol. 68 (1966), pp. 883–98; Arthur P. Wolf, "Childhood association and sexual attraction: A further test of the Westermarck hypothesis," *American Anthropologist*, vol. 72 (1970), pp. 503–15; Arthur P. Wolf, *Sexual Attraction and Childhood Association: A Chinese Brief for Edward Westermarck* (Stanford, Calif.: Stanford University Press, 1995). See also Chapter 4 of this volume.

6. See Chapter 2 of this volume and Pusey and Wolf, "Inbreeding avoidance in animals," p. 201.

7. See Leah Gavish, Joyce E. Hofmann, and Lowell L. Getz, "Sibling recognition in the prairie vole, *Microtus ochrogaster*," *Animal Behavior*, vol. 23 (1984), pp. 362–66; Maurice Greenberg and Roland Littlewood, "Post-adoption incest and phenotypic matching: Experience, personal meanings, and biosocial implications," *British Journal of Medical Psychology*, vol. 68 (1995), pp. 29–44.

8. Sigmund Freud, "Totem and Taboo," in *Complete Psychological Works*, standard edition, vol. 13 (London: Hogarth Press, 1953); Claude Lévi-Strauss, "The family," in *Man, Culture, and Society*, ed. H. L. Shapiro (London: Oxford University Press, 1956); 278; Leslie A. White, "The definition and prohibition of incest," *American Anthropologist*, vol. 50 (1948), pp. 416–35.

9. Edward Westermarck, *The History of Human Marriage*, vol. 2 (New York: Allerton Press, 1922).

10. Donald S. Sade, "Inhibition of son-mother mating among free ranging rhesus monkeys," *Science and Psychoanalysis*, vol. 12 (1968), pp. 18–38.

11. See Chapter 3 of this volume and Pusey and Wolf, "Inbreeding avoidance in animals," pp. 202–5.

12. See Hudson K. Reeve, David F. Westneat, William A. Noon, Paul Sherman,

and C. F. Aquado, "DNA 'fingerprinting' reveals high levels of inbreeding colonies of eusocial naked mole-rats," *Proceedings of the National Academy of Science, USA*, vol. 87 (1990), pp. 2496–500.

13. Greta Agren, "Incest avoidance and bonding between siblings in gerbils," *Behavioral Ecology and Sociobiology*, vol. 14 (1984), pp. 161–69; Gavish, Hofmann, and Getz, "Sibling recognition in the prairie vole"; Dustin Penn and William Potts, "MHC-disassortative mating preferences reversed by cross-fostering," *Proceedings of the Royal Society of London B*, vol. 265 (1998), pp. 1299–306; Kunio Yamazaki, Gary B. Beauchamp, D. Kupriewski, Judith Bard, L. Thomas, and Edward A. Boyce, "Familial imprinting determines H-2 selective mating preferences," *Science*, vol. 240 (1988), pp. 1331–32.

14. Gavish, Hofmann, and Getz, "Sibling recognition."

15. See Chapter 7 of this volume. Contrast with Wolf, *Sexual Attraction*, pp. 497–515.

16. Bruno Bettelheim, *Children of the Dream* (New York: Macmillan, 1969); Albert I. Rabin, *Growing Up on a Kibbutz* (New York: Springer, 1965); Shepher, "Mate selection"; Melford E. Spiro, *Children of the Kibbutz* (New York: Schocken Books, 1965); Yonina Talmon, "Mate selection in collective settlements," *American Sociological Review*, vol. 29 (1964) pp. 491–508. See Chapter 4 of this volume and Wolf, *Sexual Attraction*.

17. Shepher, "Mate selection."

18. See Chapter 4 of this volume and Wolf, *Sexual Attraction*.

19. Wolf, *Sexual Attraction*, pp. 264–375.

20. Ibid., pp. 204–13.

21. Edward O. Wilson, *Consilience: The Unity of Knowledge* (New York: Alfred E. Knopf, 1998), pp. 173–76.

22. Alfred C. Kinsey, Wardell B. Pomeroy, Clyde E. Martin, and Paul H. Gebhard, *Sexual Behavior in the Human Female* (Philadelphia: W. B. Saunders, 1953).

23. S. Kirson Weinberg, *Incest Behavior* (New York: Citadel, 1955), pp. 38–39.

24. A. M. Freedman, Howard I. Kaplan, and Benjamin J. Sadock, *Comprehensive Textbook of Psychiatry*, 2nd ed. (Baltimore: Williams and Wilkins, 1975), as cited in Herman, *Father-Daughter Incest*.

25. Diana E. H. Russell, "Intrafamilial and extrafamilial sexual abuse of female children," *Child Abuse and Neglect*, vol. 7 (1983), pp. 133–46.

26. Ibid.

27. Daniel S. Halperin, Paul Bouvier, Philip D. Jaffe, Roger-Luc Mounoud, Claus H. Pawlak, Jerome Laederach, Helene R. Wickey, and Florence Astie, "Prevalence of child sexual abuse among adolescents in Geneva: Results of a cross sectional survey," *British Medical Journal*, vol. 312 (1996), p. 1326.

28. See Russell, "Intrafamilial and extrafamilial sexual abuse," p. 135.

29. See David Finkelhor, Gerald Hotaling, I. A. Lewis, and Christine Smith, "Sexual abuse in a national survey of adult men and women: Prevalence, characteristics, and risk factors," *Child Abuse and Neglect*, vol. 14 (1990), p. 20.

30. In Finkelhor's 1980 study, 13 percent of the entire sample of 796 undergraduate students reported some type of sexual experience with a sibling (15 per-

cent of females and 10 percent of males). Most of these experiences were regarded as childhood sex play. About 25 percent of this group, or a prevalence of about 3.2 percent, had experiences regarded as exploitative because force was used or because there was a large age difference between siblings. In this study and those by Baker and Ducan and Finkelhor, Hotaling, Lewis, and Smith the precise degree of relatedness was not specified. These data may include incest involving kin outside the immediate family or involving step-parents or step-siblings. See David Finkelhor, "Sex among siblings: A survey of prevalence, variety, and effects," *Archives of Sexual Behavior*, vol. 9 (1980), pp. 171–94; Anthony W. Baker and Sylvia P. Duncan, "Child sexual abuse: A study of prevalences in Great Britain," *Child Abuse and Neglect*, 9 (1985), 457–67; and Finkelhor, Hotaling, Lewis, and Smith, "Sexual abuse in a national survey."

31. The data cited from Halperin et al.'s 1996 Geneva study are unpublished findings. I am grateful to Paul Bouvier and Daniel Halperin of the Service De Santé De La Jeunesse of Geneva, Switzerland, for their generosity in providing these data.

32. Russell, "Intrafamilial and extrafamilial sexual abuse," p. 141.

33. Baker and Duncan, "Child Sexual Abuse," p. 461.

34. Loretta M. McCarty, "Mother-child incest: Characteristics of the offender," *Child Welfare*, vol. 65 (1986), pp. 447–58.

35. For a review see L. Roybal and Jean Goodwin, "The incest pregnancy," in *Sexual Abuse: Incest Families and Their Victims*, ed. Jean Goodwin (Chicago: New Year Medical Publisher, 1989.

36. David Finkelhor, "The international epidemiology of child sexual abuse," *Child Abuse and Neglect*, vol. 18 (1949), pp. 409–17.

37. See Linda M. Williams and David Finkelhor, "Parental care-giving and incest: Test of a biosocial model," *American Journal of Orthopsychiatry*, vol. 65 (1995), pp. 101–13; and also Hilda Parker and Seymour Parker, "Father-daughter sexual abuse: An emerging perspective," *American Journal of Orthopsychiatry*, vol. 56 (1986), pp. 532–49.

38. Herman, *Father-Daughter Incest*, p. 71.

39. Hector Cavallin, "Incestuous fathers: A clinical report," *American Journal of Psychiatry*, vol. 122 (1966), pp. 1132–38; Herman, *Father-Daughter Incest*, pp. 67–125; Marisa Laviola, "Effects of older brother–younger sister incest: A study of dynamics of 17 cases," *Child Abuse and Neglect*, vol. 16 (1992), pp. 409–21; Noel Lustig, John W. Dresser, Seth W. Spellman, and Thomas B. Murray, "Incest: A family group survival pattern," *Archives of General Psychiatry*, vol. 14 (1966), pp. 31–40; Philip G. Madonna, Susan Van Scoyk, and David P. Jones, "Family interactions within incest and non-incest families," *American Journal of Psychiatry*, vol. 148 (1991), pp. 46–49; McCarty, "Mother-child incest"; Marcellina Mian, Peter Marton, Deborah LeBaron, and David Birdwistle, "Familial risk factors associated with intrafamilial and extrafamilial sexual abuse of three to five year old girls," *Canadian Journal of Psychiatry*, vol. 39 (1994), pp. 348–53; Holly Smith and Edie Israel, "Sibling incest: A study of the dynamics of 25 cases," *Child Abuse and Neglect*, vol. 11 (1987), pp. 101–8.

40. Madonna, Van Scoyk, and Jones, "Family interactions," pp. 47–48.

41. Kathleen C. Faller, "Women who sexually abuse children," *Violence and Victims*, vol. 2 (1987), pp. 263–75; Herman, *Father-Daughter Incest*, p. 76; McCarty, "Mother-child incest," p. 452; Mian, Marton, LeBaron, and Birdwistle, "Familial risk factors," p. 350; Heikki Sariola and Antti Uutela, "The prevalence and context of sexual abuse in Finland," *Child Abuse and Neglect*, vol. 20 (1996) p. 846.

42. Cavallin, "Incestuous fathers," p. 1134; Herman, *Father-Daughter Incest*, p. 78; Laviola, "Effects of older brother–younger sister incest," p. 414; Narcyz Lukianowicz, "Incest I: Paternal incest," *British Journal of Psychiatry*, vol. 120 (1972), pp. 301–13; Jane M. Rudd and Sharon D. Herzberger, "Brother-sister incest and father-daughter incest: A comparison of characteristics and consequences," *Child Abuse and Neglect*, vol. 9 (1999), pp. 915–28.

43. Herman, *Father-Daughter Incest*, p. 86; Sariola and Uutela, "Sexual abuse in Finland," p. 848.

44. Patricia Phelan, "Incest and its meaning: The perspectives of fathers and daughters," *Child Abuse and Neglect*, vol. 19 (1995), pp. 7–24.

45. See Lukianowicz, "Incest I," p. 303.

46. Herman, *Father-Daughter Incest*, pp. 80–81.

47. McCarty, "Mother-child incest"; Faller, "Women who sexually abuse children."

48. Lukianowicz, "Incest I," pp. 303–7; Lustig Dresser, Spellman, and Murray, "Incest," p. 38; Mian, Marton, LeBaron, and Birdwistle, "Familial risk factors," p. 352.

49. Baker and Duncan, "Child Sexual Abuse," p. 459; Cavallin, "Incestuous Fathers," p. 1132; Herman, *Father-Daughter Incest*, pp. 67–72.

50. Herman, *Father-Daughter Incest*, pp. 50–63.

51. Madonna, Van Scoyk, and Jones, "Family interactions," pp. 47–48; David W. Smith and Benjamin E. Saunders, "Personality characteristics of father/daughter perpetuators and non-offending mothers in incest families: Individual and dyadic analysis," *Child Abuse and Neglect*, vol. 19 (1995), pp. 607–17.

52. Mark T. Erickson, "Incest avoidance and familial bonding," *Journal of Anthropological Research*, vol. 45 (1989), pp. 267–91; Mark T Erickson, "Rethinking Oedipus: An evolutionary perspective on incest avoidance," *American Journal of Psychiatry*, vol. 150 (1993), pp. 411–16; Mark T. Erickson, "Incest avoidance: Clinical implications of the evolutionary perspective," in *Evolutionary Medicine*, ed. Wenda R. Trevathan, E. O. Smith, and James J. McKenna (New York: Oxford University Press, 1991), pp. 165–81; Arthur P. Wolf, "Westermarck redivivus," *Annual Review of Anthropology*, vol. 22 (1993), pp. 157–75; Wolf, *Sexual Attraction*, pp. 463–75.

53. Cavallin, "Incestuous fathers," p. 1134; Parker and Parker, "Father-daughter sexual abuse"; Blair Justice and Rita Justice, *Broken Taboo* (New York: Human Sciences Press, 1979), p. 63; Irving Kaufman, Alice L. Peck, and Consuela Tagiuri, "The family constellation and overt incestuous relations between father and daughter," *American Journal of Orthopsychiatry*, vol. 24 (1954), pp. 266–77; Lustig, Dresser, Spellman, and Murray, "Incest," p. 33; David L. Raphling, Bob L. Carpenter, and Allen Davis, "Incest: A genealogical

study," *Archives of General Psychiatry*, vol. 16 (1967), pp. 505–11; Mian, Marton, LeBaron, and Birdwistle, "Familial risk factors," p. 350; Patricia K. Mrazek, "The nature of incest: A review of contributing factors," in *Sexually Abused Children and Their Families*, ed. Patricia K. Mrazek and Charles H. Kempe (Elmsford, N. Y.: Pergamon Press, 1981); Williams and Finkelhor, "Parental care-giving and incest," p. 106.

54. Justice and Justice, *Broken Taboo*, p. 63; Mian, Marton, LeBaron, and Birdwistle, "Familial risk factors," p. 350; Lustig, Dresser, Spellman, and Murray, "Incest," p. 34; Mrazek, "The nature of incest."

55. McCarty, "Mother-child incest," pp. 448–50.

56. Laviola, "Effects of older brother–younger sister incest," pp. 414–15; Rudd and Herzberger, "Brother-sister incest and father-daughter incest," p. 919; Smith and Israel, "Sibling incest."

57. Rudd and Herzberger, "Brother-sister incest and father-daughter incest," p. 919; Smith and Israel, "Sibling incest."

58. Naomi A. Adler and Joseph Schultz, "Sibling incest offenders," *Child Abuse and Neglect*, vol. 19 (1995), pp. 811–19.

59. Smith and Israel, "Sibling incest," p. 104.

60. Laviola, "Effects of older brother-younger sister incest," p. 415.

61. Kaufman, Peck, and Tagiuri, "The family constellation," p. 266; Lukianowicz, "Incest I," p. 304; Phelan, "Incest and its meaning," p. 9; Sariola and Uutela, "Sexual abuse in Finland," p. 846; Gail Wyatt, "Sexual abuse of Afro-American and white-American women in childhood," *Child Abuse and Neglect*, vol. 9 (1985), p. 514.

62. Lukianowicz, "Incest I," p. 304; Herman, *Father-Daughter Incest*, p. 83; Phelan, "Incest and its meaning," p. 9.

63. Faller, "Women who sexually abuse children," p. 271.

64. Adler and Schultz, "Sibling incest offenders," p. 815.

65. Roy R. Frenzel and Reuben A. Lang, "Identifying sexual preferences in intra-familial and extra-familial child sexual abusers," *Annals of Sex Research*, vol. 2 (1989), pp. 255–75.

66. Anne E. Pusey, personal communication, September 19, 2000.

67. Gavish, Hofmann, and Getz, "Sibling recognition in the prairie vole."

68. Weinberg, *Incest Behavior*, p. 78.

69. Greenberg and Littlewood, "Post-adoption incest," pp. 29–44 (for quotes see p. 35).

70. Sarah Strickland, "A love that does not know its true name," *The Independent on Sunday* (London), January 3, 1993, p. 20.

71. Daniel Voll, "An American family," *Esquire*, July, 1998, pp. 122–45.

72. See *Time* magazine, July 2, 1979, and the *New York Times*, August 2, 1979.

73. Barbara J. Lifton, *Lost and Found: The Adoption Experience* (New York: Dial Press, 1979), p. 165.

74. Benita Eissler, *Byron: Child of Passion, Fool of Fame* (New York: Alfred Knopf, 1999).

75. Deirdre Bair, *Anaïs Nin: A Biography* (New York: Putnam and Sons, 1995), pp. 15–17.

76. Ibid., p. 170.

77. Ibid., p. 173.

78. D. W. Cory and R. E. L. Masters, *Violations of Taboo: Incest of the Great Literature of Past and Present* (New York: Julian Press, 1963); Otto Rank, *The Incest Theme in Literature and Legend*, trans. Gregory C. Richter (Baltimore: Johns Hopkins University Press, 1992).

79. To my knowledge, the only documentation of exceptions to marriage between immediate kin, associated from early life, come from ancient sources such as the reports of brother-sister marriage in Roman Egypt, circa A.D. 20–258. A degree of skepticism is warranted. Direct contact with these couples is, of course, impossible. The importance of direct contact or observation has been made abundantly apparent in studies of incest avoidance. It was thought that incest was common in nature until careful observational studies revealed otherwise. In Taiwan a number of couples who had married in the minor fashion told Wolf that they had not been sexually active despite years of marriage. Their children were conceived in extramarital affairs but registered in the family name. It is impossible to rule out an explanation of this sort for suggestive findings from antiquity. See, for example, Keith Hopkins, "Brother-sister marriage in Roman Egypt," *Comparative studies in Society and History*, vol. 22 (1980), pp. 303–47, and Chapter 5 in this volume. See also Wolf, *Sexual Attraction*, pp. 427–29.

80. Kathleen Harrison, *The Kiss: A Memoir* (New York: Random House, 1997), p. 56.

81. See Paul W. Sherman and Warren G. Holmes, "Kin-recognition: Issues and evidence," in *Experimental Behavioral Ecology and Sociobiology*, ed. Bert Hölldobler and Martin Lindauer (Sunderland, Mass.: Sinauer, 1985).

82. See Richard H. Porter, "Mutual mother-infant-recognition in humans," in *Kin Recognition*, ed. P. G. Hopper (Cambridge: Cambridge University Press, 1991).

83. Marsha Kaitz, Shimon Shuri, Shai Danziger, Ziva Herskko, and Arthur I. Eidelman, "Fathers can also recognize their newborns by touch," *Infant Behavior and Development*, vol. 17 (1994), pp. 205–7.

84. Marsha Kaitz, A. Good, A. M. Rokem, and A. I. Eidelman, "Mothers' recognition of their newborn by olfactory cues," *Development Psychology*, vol. 20 (1988), pp. 582–91.

85. Patrick Bateson, "Sexual imprinting and optimal outbreeding," *Nature*, vol. 273 (1978), 659–60.

86. Greenberg and Littlewood, "Post-adoption incest," p. 41–43.

87. William D. Hamilton, "Genetical evolution of social behavior," *Journal of Theoretical Biology*, vol. 7 (1964), pp. 1–52; John Maynard Smith, "Group selection and kin selection," *Nature*, vol. 201 (1964), pp. 1145–47; George C. Williams, *Adaptation and Natural Selection: A Critique of Some Current Evolutionary Thought* (Princeton, N.J.: Princeton University Press, 1966).

88. George C. Williams and Randolph M. Nesse, "Dawn of Darwinian medicine," *Quarterly Review of Biology*, vol. 66 (1991), pp. 1–22.

89. Paul W. Sherman, "Nepotism and the evolution of alarm calls," *Science*, vol. 197 (1977), pp. 1246–53; Evelyn B. Hangii and Ronald J. Schusterman, "Kin

recognition in captive California sea lions," *Journal of Comparative Psychology*, vol. 104 (1990), pp. 368–72; D. R. Meikle and S. H. Vessey, "Nepotism among rhesus monkey brothers," *Nature*, vol. 294 (1984), pp. 160–61.

90. Robert Axelrod and William D. Hamilton, "The evolution of cooperation," *Science*, vol. 211 (1981), pp. 1390–96.

91. John Bowlby, *Attachment and Loss*, vol. I: *Attachment* (New York: Basic Books, 1969) p. 305.

92. Patrick Bateson, "How do sensitive periods arise and what are they for?" *Animal Behavior*, vol. 27 (1979), pp. 470–86.

93. Sherman and Holmes, "Kin-recognition."

94. Wolf, "Westermarck redivivus," p. 168; Wolf, *Sexual Attraction*, pp. 463–75; Erickson, "Incest avoidance and familial bonding," pp. 278–80; Erickson, "Rethinking Oedipus," p. 413; Erickson, "Incest Avoidance."

95. Wolf, *Sexual Attraction*, p. 471.

96. Bill Amos, Christian Schlotterer, and Deihard Tautz, "Social structure of pilot whales revealed by DNA profiling," *Science*, vol. 260 (1993), pp. 670–72.

97. Frans de Waal and Frans Lanting, *Bonobo: The Forgotten Ape* (Berkeley: University of California Press, 1997), pp. 115–17.

98. Francois Jacob, "Evolution and tinkering," *Science*, vol. 196 (1977), pp. 1161–66.

99. Mary D. S. Ainsworth, *Infancy in Uganda: Infant Care in the Growth of Love* (Baltimore: Johns Hopkins University Press, 1967); Russell A. Isabella, "Origins of attachment: Maternal interactive behavior across the first year," *Child Development*, vol. 64 (1993), pp. 605–21; Russell A. Isabella and Jay Belsky, "Interactional synchrony and the origins of infant-mother attachment," *Child Development*, vol. 62 (1991), pp. 373–84; Philip B. Smith and David R. Pederson, "Maternal sensitivity and patterns of infant-mother attachment," *Child Development*, vol. 59 (1988), pp. 1097–101.

100. Howard Steele, Miriam Steele, and Peter Fonagy, "Association among attachment classifications of mothers, fathers, and their infants," *Child Development*, vol. 67 (1996), pp. 541–55. See also Marius H. van Ijzendoorn, "Adult attachment representations, parental responsiveness, and infant attachment: A meta-analysis on the predictive validity of the adult attachment interview," *Psychological Bulletin*, vol. 117 (1995), pp. 387–403.

101. Robin Fox, "Sibling incest," *British Journal of Sociology*, vol. 13 (1962), pp. 128–50.

102. Meyer Fortes, *Web of Kinship Among the Tallensi* (London: Oxford University Press, 1957), p. 251; Kinsey, Pomeroy, Martin, and Gebhard, *Sexual Behavior in the Human Female*, p. 104.

103. William J. Demarest, "Incest avoidance among human and nonhuman primates," in *Primate Social Behavior*, ed. Susan Chevalier-Skolnikoff and E. F. Poirer (New York: Garland Press, 1977), pp. 323–42.

104. Frank A. Beach, "Instinctive behavior: Reproductive activities," in *Handbook of Experimental Psychology*, ed. S. S. Stevens (New York: Wiley, 1951), p. 408.

105. Howard A. Bacal and Kenneth M. Newman, *Theories of Object Rela-

tions to Self Psychology (New York: Columbia University Press, 1990), p. 12; George A. DeVos, "Affective dissonance and primary socialization: Implications for a theory of incest avoidance," *Ethos*, vol. 3 (1975), pp. 165–82; Takeo Doi, "The concept of amae and its psychoanalytic implications," *International Review of Psychoanalysis*, vol. 16 (1989), pp. 349–54; Sandor Ferenczi, "Confusion of tongues between adult and the child," *International Journal of Psychoanalysis*, vol. 30 (1949), pp. 225–30; Heinz Kohut, *How Does Analysis Cure?* ed. A. Goldberg and P. Stepansky (Chicago: University of Chicago Press, 1984), p. 3–12; Ian D. Suttie, *Origins of Love and Hate* (London: Kegan Paul, 1935), pp. 80–96.

106. Erickson, "Incest avoidance and familial bonding," pp. 278–80; Erickson, "Rethinking Oedipus," p. 413; Wolf, "Westermarck redivivus," p. 168; Wolf, *Sexual Attraction*, pp. 463–75.

107. See Rabin, *Growing Up on a Kibbutz*, p. 23.

108. See Shepher, "Mate selection," p. 296.

109. Allen L. Sroufe and Mary J. Ward, "Seductive behaviors of mothers of toddlers: Occurrence, correlates, and family origins," *Child Development*, vol. 51 (1980), pp. 1222–29.

110. Clare Haynes-Seman and Richard Krugman, "Sexualized attention: Normal interaction or precursor to sexual abuse?" *American Journal of Orthopsychiatry*, vol. 59 (1989), pp. 238–45.

111. Herman, *Father-Daughter Incest*, pp. 109–25.

112. Mary Main and Ruth Goldwyn, *Adult Attachment Classification System: Version 5*. Unpublished manuscript, University of California, Berkeley, 1991.

113. Paul Chodoff, "A critique of Freud's theory of infantile sexuality," *American Journal of Psychiatry*, vol. 123 (1966), pp. 507–18.

114. See Chapter 3 of this volume.

115. de Waal and Lanting, *Bonobo*, p. 117.

116. Allen L. Sroufe, Christopher Bennett, Michelle Englund, and Shmuel Shulman, "The significance of gender boundaries in preadolescence: Contemporary correlates and antecedents of boundary violation and maintenance," *Child Development*, vol. 64 (1993), pp. 455–676.

117. Bateson, "Optimal outbreeding."

118. Patrick Bateson, "Preferences for cousins in Japanese quail," *Nature*, vol. 295 (1982), pp. 236–37.

119. Claus Wedekind, Thomas Seebeck, Florence Bettens, and Alexander J. Paepke, "MHC dependent mate preferences in humans," *Proceedings of the Royal Society of London B*, vol. 260 (1995), pp. 245–49.

120. Gilbert Herdt and Martha McClintock, "The magical age of 10," *Archives of Sexual Behavior*, vol. 29 (2000), pp. 587–606.

121. See Chapter 4 of this volume; Shepher, "Mate selection"; and Greenberg and Littlewood, "Post-adoption incest," pp. 29–44.

122. See Gavish, Hofmann, and Getz, "Sibling recognition in the prairie vole."

123. See Adler and Schultz, "Sibling incest offenders"; McCarty, "Mother-child incest"; Williams and Finkelhor, "Parental care-giving and incest."

124. Anne E. Pusey, personal communication, September 19, 2000.

125. S. Boyd Eaton, S. Boyd Eaton III, and Melvin J. Konner, "Paleolithic nu-

trition revisited," in *Evolutionary Medicine*, ed. Wenda R. Trevathan, E. O. Smith, and James J. McKenna (New York: Oxford University Press, 1999), pp. 313–32.

126. Wenda R. Trevathan, "Evolutionary obstetrics," in *Evolutionary Medicine*, ed. Wenda R. Trevathan, E. O. Smith, and James J. McKenna (New York: Oxford University Press, 1999), pp. 183–208.

127. Susan M. O'Connor, Peter M. Vietze, John B. Hopkins, and William A. Altemeier, "Post-partum extended maternal-infant contact: Subsequent mothering and child health," *Pediatric Research*, vol. 11 (1977).

128. Michael Lamb, "Early mother-neonate contact and the mother-child relationship," *Journal of Child Psychology and Psychiatry*, vol. 24 (1983), pp. 487–94.

129. Dario Maestripieri, "Is there mother-infant bonding in primates?" *Development Review*, vol. 21 (2001), pp. 93–201.

130. Ibid., p. 113; Eva Nissen, Gunilla Lilia, Ann-Marie Widstrom, and Kerstin Uvnas-Moberg, "Elevation of oxytocin levels early in post-partum women," *Acta Obstetricia et Gynecologia Scandanavica*, vol. 74 (1995), pp. 530–33.

131. Katherine A. Dettwyler, "A time to wean," *Breastfeeding Abstracts*, vol. 14 (1994), pp. 3–4.

132. Ronald G. Barr, "Infant crying behavior and colic: An interpretation in evolutionary perspective," in *Evolutionary Medicine*, ed. Wenda R. Trevathan, E. O. Smith, and James J. McKenna (New York: Oxford University Press, 1999), pp. 27–52.

133. James J. McKenna, Sarah Mosko, and Chris Richard, "Breastfeeding and mother-infant co-sleeping in relation to SIDS prevention," in *Evolutionary Medicine*, ed. Wenda R. Trevathan, E. O. Smith, and James J. McKenna (New York: Oxford University Press, 1999), pp. 53–74; James J. McKenna, Evelyn B. Thoman, Thomas F. Anders, Abraham Sadeh, Vicki L. Schechtman, and Steven F. Glotzbach, "Infant-parent co-sleeping in an evolutionary perspective: Implications for understanding infant sleep development and sudden infant death syndrome," *Sleep*, vol. 16 (1993), pp. 263–82.

134. John Forbes and David S. Weiss, Cosleeping habits of military children," *Military Medicine*, vol. 157 (1992), pp. 196–200.

135. Parker and Parker, "Father-daughter sexual abuse."

136. Williams and Finkelhor, "Parental care-giving and incest."

10 The Incest Taboo as Darwinian Natural Right

Larry Arnhart

In the twenty-first century, we will see great advances in the biological understanding of human nature. This will force us to think about whether biological science can explain that most distinctive trait of our humanity—our moral sense of right and wrong. Some people will argue that our moral experience transcends our biological nature. Others will argue that we should be able to explain our morality as an expression of our biological nature. How we decide that debate might be decisively influenced by whether we accept or reject Edward Westermarck's Darwinian theory of the incest taboo as a natural expression of human moral emotions.

Westermarck contended that incest is wrong because most human beings are naturally inclined to learn an emotional aversion to incest. The incest taboo varies in its details because different societies have different systems of kinship. But generally across all societies incestuous relations with one's siblings, parents, or children is regarded as wrong because it elicits moral emotions of disapproval. Westermarck inferred that since close inbreeding tended to have deleterious effects on offspring that would lower their fitness for survival and reproduction, natural selection would have favored the natural tendency to feel a sexual aversion toward those with whom one had been reared in early childhood, and this emotional aversion toward incest would tend to be expressed as social disapproval through an incest taboo. Consequently, there is no universal, cosmic imperative of reason that dictates that incest is wrong. Rather, incest is wrong for a particular species of animal that has a natural propensity to develop strong moral emotions against incest. Even within the human species, there is variation in this propensity, such that some individuals do not feel an aversion to incest, but they will be treated by the rest of us as moral strangers. Westermarck thought this illustrated the general character of ethics as rooted in the moral emotions of the human animal as shaped by natural selection in evolutionary history.[1]

The importance of Westermarck's theory for biological explanations of ethics is evident in Edward O. Wilson's work and the continuing controversy over his work. Harvard University Press recently published the twenty-fifth anniversary edition of Wilson's *Sociobiology*. From its first publication in 1975, that book has stirred a dispute over his claim that ethics is rooted in human biology. Our deepest intuitions of right and wrong, he asserted on the first page of the book, are guided by the emotional control centers of the brain, which evolved by natural selection to help the human animal exploit opportunities and avoid threats in the natural environment.[2] In 1998 Wilson's book *Consilience* renewed the controversy as he continued to argue for explaining ethics through the biology of the moral sentiments.[3] In recent decades, the dispute over Wilson's intellectual project has become one of the most prominent public debates over the application of modern science to the study of human social life.[4] Since morality is ultimately rooted in the moral sentiments of human nature, Wilson claims, a natural science of morality would require a biology of the moral sentiments.[5] Throughout his writings, Wilson has used Westermarck's Darwinian theory of the incest taboo as the prime example of how biology can explain the moral sentiments.[6] Human nature is not a product of genes alone or of culture alone, Wilson insists. Rather, human nature is constituted by "the epigenetic rules, the hereditary regularities of mental development that bias cultural evolution in one direction as opposed to another, and thus connect the genes to culture."[7] The biology of the moral sentiments would be the study of the "epigenetic rules" of moral experience as shaped by the complex interaction of genetic propensities and cultural learning. And for Wilson, Westermarck's theory of the incest taboo is the best example of this.

In the debate over the ultimate grounds of ethics, Wilson sees two fundamentally opposed positions. The "transcendentalists" claim that ethics is rooted in absolute standards that exist outside the human mind, while the "empiricists" claim that ethics is rooted in natural human inclinations.[8] Wilson defends the empiricist position, which he thinks can be traced back to Aristotle's *Nicomachean Ethics*. He rejects the transcendentalist position, which he thinks is evident in the ethical writings of philosophers like Immanuel Kant and John Rawls.[9]

It is remarkable that so few people accept Wilson's biological view of ethics as rooted in natural moral sentiments. It is hardly surprising that his most fervent critics reject his account of ethics as confirming their charge of crude biological reductionism and determinism. But it is surprising that even those who generally agree with him about applying Darwinian theory to human behavior reject his Darwinian view of ethics. All of Wilson's intellectual enemies and almost all of his intellectual friends seem to agree that there is an absolute separation between natural facts and moral values, between *is* and *ought*, so that a biological science of ethics would commit the "naturalistic fallacy" by inferring a moral *ought* from a natural *is*.

Professional philosophers—like Thomas Nagel, for example—reject any Darwinian account of ethics for failing to see that ethics is "an autonomous theoretical subject" that manifests the "moral point of view" as attained by the exercise of human reason in transcending human biology.[10] Even the proponents of "evolutionary psychology," who are generally supportive of Wilson's position, reject any biological science of ethics as violating the fact-value dichotomy. David Buss, for example, in his Darwinian account of human mating, warns against the "naturalistic fallacy," which, he says, "confuses a scientific description of human behavior with a moral prescription for that behavior." He insists that "judgments of what should exist rest with people's value systems, not with science."[11] Steven Pinker explains almost everything about human behavior in Darwinian terms, but even he concludes that there is no scientific explanation for ethics. "How did *ought* emerge from a universe of particles and planets, genes and bodies?" Natural science cannot answer that question, Pinker asserts, because our human ability to reason from a moral *ought* transcends our natural inclinations.[12] When Wilson delivered the keynote address to the 1996 annual meeting of the Human Behavior and Evolution Society, which is the principal organization for scholars applying Darwinian views of human nature to the study of human social behavior, he spoke of the need for a unity of the sciences that would promote a biological science of ethics. Many of those in the audience were shocked by his denial of the fact-value dichotomy and his refusal to accept the transcendent autonomy of ethics.[13] It seems that almost everyone is a "transcendentalist" in viewing ethics as manifesting transcendent values, and almost no one is persuaded by Wilson's "empiricist" view of ethics as rooted in natural moral emotions. Almost everyone agrees that because of the dichotomy between *is* and *ought*, reasoning about the moral *ought* belongs to a transcendent realm of human freedom beyond the empirical world of nature.

When I first read Wilson's *Sociobiology* in 1975, I too rejected his biological explanation of ethics as being too crudely reductionistic in its appeal to mere emotion as the ultimate ground of ethical experience. At the time, I was a graduate student at the University of Chicago writing a dissertation on Aristotle's *Rhetoric*, and I noticed that Aristotle invoked a natural moral sense as expressed in moral emotions such as anger, indignation, shame, kindness, and pity, which made me think that Wilson's reliance on the moral emotions might be more defensible that I had initially believed.[14] I also became interested in Aristotle's biological writing and in how his biological reasoning influenced his ethical philosophy. Like Wilson, Aristotle explained the natural sociality of human beings by comparing them with other social animals such as the social insects. When Aristotle spoke of "natural right"—natural standards of right and wrong—he appealed to the biological propensities of human nature. I also saw that Aristotelian philosophers like Thomas Aquinas spoke of "natural right" or "natural law" as "that which nature has taught all animals." Eventually, I changed my mind about Wilson's argument and concluded that his Darwinian explanation of ethics could be defended as a modern biological restatement of a tradition of ethical naturalism that began with Aristotle.

Now, I argue that a Darwinian science of human nature can explain ethics as conforming to what I call "Darwinian natural right."[15] I also agree with Wilson that Westermarck's theory of the incest taboo is the best illustration of how such a natural science of ethics might work. A Darwinian view of human nature can support a form of ethical naturalism that is compatible with the philosophic tradition of reasoning about "natural right" or "natural law," a tradition that stretches from Plato and Aristotle in ancient Greece to Thomas Aquinas in the Middle Ages and then to David Hume and Adam Smith in the early modern world. Furthermore, throughout that tradition, the incest taboo is one of the prime examples of how ethics might be rooted in the moral emotions of human nature. The great achievement of Westermarck's theory of incest is how it sustains that philosophic tradition by providing a Darwinian explanation of the natural moral emotions associated with the incest taboo. Wilson is right, therefore, in seeing that recent research confirming Westermarck's theory provides powerful support for a naturalistic view of ethics as founded in the moral sentiments of human nature.

The Natural Right Tradition

Among the Greek sophists, it was common to contrast *physis* and *nomos*, "nature" and "convention." (Today we would speak of this as the dichotomy between "nature" and "nurture.") The sophists often asserted that the social norms of human conduct are rooted in "convention" rather than "nature." Aristotle thought Callicles in Plato's *Gorgias* was speaking for the sophists generally in arguing that "nature and convention are opposites, and justice is a noble thing according to convention, but not so according to nature."[16] By nature, the sophists seemed to say, human beings are too selfish and competitive to live together in civilized communities, and therefore social order requires the invention of social conventions or laws that repress the otherwise unruly desires of human nature.[17] If there is any natural standard of right at all, Callicles declared, it would dictate the rule of the stronger over the weaker, but this is exactly what conventional morality denies.

Against the sophists, Plato, Aristotle, and the other Socratic philosophers argued that justice was not merely conventional, because one could discover norms that were just or right by nature. The Socratic philosophers thus began a tradition of reasoning about "natural right" or "natural law." As the very term "natural law" suggests, this tradition tries to overcome the sophistic antinomy of "nature" versus "convention," because the thought is that some conventions are "according to nature" and others are "contrary to nature." But how are we to understand the complex interaction of nature and convention? And how are we to distinguish the natural conventions from the unnatural conventions? If social order is rooted in a universal human nature, how can we explain the apparent diversity of social conventions across different societies? I believe that a Darwinian understanding of the biological nature of social order helps to answer these questions, and thus it vindicates the Socratic tradition of ethical naturalism against the sophistic tradition of ethical conventionalism.

In Plato's *Laws*, Socrates speaks of the avoidance of incest as an "unwritten law" that is so strong that "among the many there isn't the slightest desire for this sort of intercourse." The strength of this law arises from the fact that everyone from the moment of birth hears incest condemned as "hateful to the gods and the most shameful of shameful things." Plato leaves it unclear, however, why this sacred taboo arises in the first place. The discussion of incest avoidance in Plato's dialogue arises in the context of devising laws for a good city that will be "according to nature." But there is no explicit discussion of whether—and if so, how—the law of incest avoidance might have natural causes.[18] In Xenophon's *Memorabilia*, however, Socrates identifies the "unwritten laws" legislated by the gods as laws that could not be disobeyed without natural penalties. He speaks of the incest taboo as one of those "unwritten laws," because those committing incest would tend to produce defective offspring.[19]

Aristotle developed the biological basis of this natural law in his biological writings. Even his works of ethical and political philosophy show a biological view of human nature, so that some scholars have concluded that for Aristotle, "ethics and politics are in a way biological sciences."[20] The sophists had argued that all justice was conventional and not natural, because while nature is invariable, the rules of justice are variable. But Aristotle saw that for biological phenomena, natural law is that which happens "for the most part" but not always. So, for example, there is a natural propensity among most people for the right hand to be stronger, although some people are left-handed, and some can be habituated to be ambidextrous. Similarly, what is naturally right conforms to the natural propensities or inclinations of human beings, but this will vary according to the variable circumstances of action.[21]

Thus did the Socratic philosophers of ancient Greece start a tradition of reasoning about "unwritten laws" that might be "according to nature," in which the incest taboo was one example of a natural propensity of human biological nature that was enforced by the natural penalties of disobedience. The idea of "natural law" was elaborated by Stoic philosophers in ancient Rome. Then, in the Middle Ages, Thomas Aquinas developed the idea in its fullest form. Wilson dismisses the Thomistic tradition of natural law as "transcendentalist," because the natural law is viewed as expressing God's will, and thus it is independent of the human mind.[22] But I think Wilson is mistaken. Aquinas's natural law is actually an "empiricist" position, at least insofar as Aquinas speaks of natural law as rooted in natural inclinations that are instinctive to human beings. Although Aquinas believes that God is the ultimate cause of everything in nature, the natural inclinations that constitute the natural law can be known by natural human experience even without religious faith.

Aquinas condemns incest as contrary to the "natural law" (*lex naturalis*) or "natural right" (*ius naturale*) that governs sexual mating and familial bonding. Aquinas indicates the biological character of the natural moral law when he quotes the ancient Roman jurist Ulpian as declaring that "natural right is that which nature has taught all animals." Natural law is rooted in "natural inclination" (*inclinatio naturae*) or "natural instinct" (*instinctu naturae*), and therefore "all those things to which a human being has a natural inclination, reason naturally apprehends as good, and consequently as objects to be pursued, and the contraries of these as evil and to be avoided." Natural law for human beings requires reason, but pure reason could not move us to act if it were not linked somehow to inclination or instinct (that is, some affective or emotional impulse) as the motivation to action.

Aquinas believes that human beings, like some other animals, are naturally inclined to sexual union and parental care.[23] Marriage belongs to natural law because it serves two natural inclinations—the care of offspring and conjugal bonding. Aquinas can judge the naturalness of marital arrangements, therefore, by how well they promote those two natural ends.[24] He speaks of the human disposition to marriage as a "natural instinct of the human species."[25] The primary natural end of marriage is to secure the parental care of children. The secondary natural end is to secure the conjugal bonding of male and female for a sexual division of labor in the household. Among some animals, Aquinas observes, the female can care properly for her offspring on her own, and thus there is no natural need for any enduring bond between male and female. But for those animals whose offspring require care from both parents, nature implants an inclination for male and female to stay together to provide the necessary parental care.[26] Just as is the case for those animals whose offspring could not survive or develop normally without parental care, human offspring depend upon parents for their existence, their nourishment, and their education. To secure this natural end, nature instills in human beings, as it does in other animals, natural desires for sexual coupling and parental care. Even if they do not have children, however, men and women naturally desire marital union because, not being self-sufficient, they seek the conjugal friendship of husband and wife sharing in household life.

Marriage as constituted by customary or legal rules is uniquely human, because such rules require a cognitive capacity for conceptual reasoning that no other animals have. But even so, such rules provide formal structure to natural desires that are ultimately rooted in the animal nature of human beings.

Aquinas gives at least five reasons why incestuous marriage would be contrary to natural law.[27] First, as Pope Gregory I said, "we have learned by experience that the children of such a union cannot thrive." (William Durham cites this statement from Pope Gregory as evidence that the Catholic prohibition of incest was originally based on some recognition of the physical abnormalities that come from inbreeding.)[28] Second, incest would disrupt familial relationships by impeding the reverence that children owe to their parents. Third, incest would disrupt the family by widening the range of sexual desire to include all of the kin within the household. Fourth, incest would discourage the friendly alliances that come from marrying out of one's own group. Fifth, human beings have a "natural abhorrence" of incest, an abhorrence that is even shown by some other animals. To support this last point, Aquinas cites Aristotle's claim that among camels and horses, sons naturally abhor copulation with their mothers.[29] Thus does Aquinas anticipate in some manner most of the theories developed by modern scholars to explain the universality of the incest taboo.[30]

Hobbes and Smith

The modern break with the Aristotelian and Thomistic tradition of natural law as rooted in the biology of human nature began in the seventeenth century with Thomas Hobbes. Aristotle and Aquinas had claimed that human beings are by nature social and political animals. Hobbes denied this claim and asserted that social and political order is an utterly artificial human construction. For Aristotle and Aquinas, as belonging to the Socratic tradition of ethical naturalism, moral and political order was rooted in biological nature. But for Hobbes, as belonging to the sophistic tradition of ethical conventionalism, social order required a conquest of nature by which human beings transcend their animal nature. What Hobbes identified as the "laws of nature" that should govern human social conduct were actually "laws of reason" by which human beings contrive by rational artifice to escape the disorder that ensues from following their natural inclinations.[31]

Hobbes assumes a radical separation between animal societies as founded on natural instinct and human societies as founded on social learning. Human beings cannot be political animals by nature, Hobbes believes, because "man is made fit for society not by nature but by education."[32] Against Aristotle and Aquinas, Hobbes argued that this dependence of human social order on artifice and learning meant that human beings were not at all like the naturally social animals (such as bees and ants).[33] Despite the monism of Hobbes's materialism, in which he seems to think everything is ultimately reducible to matter in motion, his political teaching presupposes a dualistic opposition between animal nature and human will: in creating political order, human beings transcend and conquer nature.[34]

This Hobbesian dualism is developed by Immanuel Kant in the eighteenth century in formulating the modern concept of culture.[35] Culture becomes that uniquely human realm of artifice in which human beings escape their natural animality to express their rational humanity as the only beings who have a "supersensible faculty" for moral will. Through culture, human beings free themselves from the laws of nature. (This Kantian notion of culture as a self-contained, autonomous sphere of uniquely human meaning eventually became the fundamental idea for cultural anthropology.)[36]

Opposing the Hobbesian claim that human beings were naturally asocial and amoral, Francis Hutcheson and other Scottish philosophers of the eighteenth century argued that human beings were endowed with the natural instincts of social animals, and this natural sociality supported a natural moral law as expressed in the natural moral sense. Hutcheson's theory of the moral sense revives the Thomistic conception of natural law as founded in the inclinations or instincts of human nature.[37]

Against Hobbes, Lord Shaftesbury asserted that there was a natural moral sense. Bernard Mandeville then responded to Shaftesbury by contending that morality was a matter of custom or convention. To illustrate this, he claimed that incestuous marriages are customary in some societies, and "there is nothing in nature repugnant against them, but what is built upon mode and custom."[38] Challenging this claim, Hutcheson insisted that the abhorrence of incest did indeed express a natural moral sense.[39]

As a student of Hutcheson's at the University of Glasgow, Adam Smith developed a similar position.[40] Smith agreed with Aquinas in condemning incestuous marriage as "shocking and contrary to nature."[41] It is contrary to nature, Smith explained, because the natural affections of familial attachment between parents and children or between siblings is contrary to the natural affections of sexual mating between husband and wife. There is, then, a natural tendency for most human beings to feel incest to be "shocking and abominable." (Similarly, proponents of Westermarck's theory of incest avoidance have argued that the "familial attraction" that bonds parents and children is evolutionarily distinct from the "sexual attraction" that bonds husband and wife, so that secure familial bonding excludes sexual bonding.[42])

But while the marriage of parent to a child or of siblings to one another would be universally contrary to nature, Smith believed, the other rules for avoiding incest can vary depending on the variable rules of kinship as determined by custom. For example, prohibiting a man from marrying his deceased wife's sister, because the wife's sister is considered to be the husband's sister, is a rule of custom rather than of nature. (The passage in 1907 of a British law allowing a man to marry his deceased wife's sister was preceded by sixty-five years of intense Parliamentary debate.[43]) Thus, the natural propensity to abhor incest will be most strongly expressed in response to relations within the immediate family circle of parents and children, but there will be variation in how far the incest taboo extends beyond the nuclear family.

While Smith spoke repeatedly of nature as instilling those moral sentiments that would promote the survival and propagation of human beings as social animals, he could not explain exactly how it was that nature could shape the human animal in this way. Such an explanation was later provided by Charles Darwin and elaborated by Edward Westermarck.

Darwin

Adopting Smith's theory of the natural moral sentiments, Darwin then went beyond Smith in showing how the moral sense could have arisen in human nature as a product of natural selection.[44] Darwin agreed with Kant and other writers "who maintain that of all the differences between man and the lower animals, the moral sense or conscience is by far the most important." But while Kant had written about the human sense of duty or "ought" as showing us "man as belonging to two worlds"—the empirical world of natural causes and the transcendent world of moral freedom—Darwin suggested that human morality could be studied "exclusively from the side of natural history."[45] In developing his evolutionary theory of morality, Darwin adopted a Smithian naturalism rather than a Kantian dualism. Like Smith, Darwin saw nature as the comprehensive whole of which human beings are a part. And thus he rejected Kant's dualistic separation between the "phenomenal" realm of causal nature and the "noumenal" realm of human freedom.

Like Aquinas and Smith, Darwin recognized the abhorrence of incest as a universal trait of human societies. And although he did not elaborate his explanation, he suggested that avoiding incest would have the advantage of preventing the bad effects of interbreeding in the same family.[46] He speculated that the feeling of sexual aversion toward those with whom one had been raised in early childhood might have been favored by natural selection in human evolutionary history.[47] This is the idea subsequently developed by Westermarck.

Westermarck

Westermarck first won international recognition with the publication in 1889 of his *History of Human Marriage*, a massive survey of the subject, in which he explained the desires for family life and sexual mating as founded in moral emotions that had been shaped by natural selection as part of the biological nature of human beings. In later writings—such as *The Origin and Development of the Moral Ideas* (1906) and *Ethical Relativity* (1932)—he elaborated a general theory of ethics as an expression of the moral emotions.[48] He combined ideas from Smith and Darwin with the comparative data of social behavior as collected by anthropologists and sociologists to defend a naturalistic theory of ethics as rooted in the natural moral sentiments. He thus provided an empirically grounded scientific theory of human nature supporting the tradition of natural law reasoning.[49]

In agreement with Aquinas, Smith, and Darwin, Westermarck argues that human marriage is natural because it satisfies some of the deepest inclinations of human nature. Against those who believe that primitive human beings lived in a state of complete promiscuity with no enduring tie between male and female, Westermarck contends that conjugal bonding has always been a natural instinct for human beings, because it was favored by natural selection to provide parental care for offspring that could not survive without such care.[50] This natural inclination to conjugal bonding gave rise to habits, customs, and institutions that sanctioned marriage as an enduring union of parents and children. As social animals with a highly developed intellect, human beings feel moral indignation toward men who abandon their wives and children, and this moral emotion of disapproval is expressed in customary and legal rules that enforce the duties and rights of spouses, parents, and children.

Much of Westermarck's work on human marriage is about the great variability in marital practices across different societies and different historical periods—ranging from monogamy to polygyny to polyandry. Even in this variation, however, he sees regularities that manifest the universal nature of human marital emotions, and they are the same regularities seen by Aquinas, Smith, and Darwin.[51] Monogamy is practiced in all societies, and in some it is the only permissible form of marriage. Monogamy is universal because it satisfies the human instincts for conjugal bonding and parental care. Although polygyny has been common in many societies, because many men have a natural desire for multiple mates, the co-wives are naturally inclined to sexual jealousy, which creates conflicts that are difficult to manage. Polyandry is the rarest form of marriage because the intense jealousy of men makes it almost impossible for them to share a wife. When polyandry is practiced, it seems to be a response to unusual circumstances, such as a low population of women in proportion to men, or harsh economic conditions that force brothers to share a wife so that their family's property is not divided.[52]

The most famous part of Westermarck's study of marriage is his theory of the incest taboo.[53] Like Plato, Aquinas, Smith, and Darwin, Westermarck sees that incest is almost universally condemned as a morally abhorrent violation of nature. All societies prohibit mothers from marrying their sons, and fathers from marrying their daughters. And with few exceptions, all societies prohibit marriages of brothers and sisters who are children of the same parents. Westermarck observes that the many theories offered to explain this are unsatisfactory. As the best summary of his reasoning, the following passage from Westermarck's *Ethical Relativity* deserves to be quoted at length.

The theories in question imply that the home is kept free from incestuous intercourse by law, custom, or education. But even if social prohibitions might prevent unions between the nearest relatives, they could not prevent the desire for such unions. The sexual instinct can hardly be changed by prescriptions; I doubt whether all laws against homosexual intercourse, even the most draconic, have ever been able to extinguish the peculiar desire of anyone born with homosexual tendencies. Nevertheless, our laws against incest are scarcely felt as a restraint upon individual feelings. And the simple reason for this is that in normal cases there is no desire for the acts which they forbid. Generally speaking, there is a remarkable absence of erotic feelings between persons living closely together from childhood; among the lower animals, also, there are indications that the pairing instinct fails to be stimulated by companions and seeks strangers for its gratifications. . . . Plato showed a sharper eye for the problem of incest in his observation that an unwritten law defends as sufficiently as possible parents from incestuous intercourse with their children and brothers from intercourse with their sisters, and that the thought of such a thing does not enter at all into the minds of most of them.

Sexual indifference, however, is not by itself sufficient to account for exogamous prohibitions. But such indifference is very generally combined with sexual aversion when the act is thought of; indeed, I believe that this is normally the case whenever the idea of sexual intercourse occupies the mind with sufficient intensity and a desire fails to appear. . . . Aversions which are generally felt readily lead to moral disapproval and prohibitory customs and laws. This I take to be the fundamental cause of the exogamous prohibitions. Persons who have been living together from childhood are as a rule near relatives. Hence their aversion to sexual relations with one another displays itself in custom and law as a prohibition of intercourse between near kin.[54]

Thus, like Plato and the other Socratic philosophers, Westermarck does not think the incest taboo can be explained simply as a product of "law, custom, or education," because "in normal cases" there is a natural aversion to incest that constitutes "an unwritten law." This natural aversion is then expressed as a legal or customary prohibition against incest. And yet this natural aversion and its expression in law and custom are not natural necessities that hold in every case but natural propensities that hold in most cases. The incest taboo, like any moral rule, is a generalization of natural emotions that hold sway "normally" in the minds of "most" people in response to circumstances that occur "as a rule." As Aristotle argued, what is naturally right is variable, but it is still natural insofar as it expresses natural propensities of the human animal diversely expressed in human custom and law.

Westermarck's Darwinian theory for explaining this can be stated in three propositions.[55] First, inbreeding tends to produce physical and mental deficiencies in the offspring that lower their fitness in the Darwinian struggle for existence. Second, as a result of the deleterious effects of inbreeding, natural selection has favored the mental disposition to feel an aversion to sexual mating with those with whom one has been intimately associated from early childhood. Third, this natural aversion to incest has inclined human beings to feel moral disapproval for incest, and this moral emotion has been expressed culturally as an incest taboo.

Westermarck's view of incest illustrates his general account of ethics as rooted in natural emotions shaped by natural selection in human evolutionary history. The avoidance of incest works through an emotional aversion favored by natural selection. Because this emotion tends to be shared by most human beings, it gives rise to moral emotions of disapproval that are expressed in customary and legal rules that prohibit incest. These customary and legal rules are culturally variable in their specific details, but the cultural rules are grounded in an emotional propensity of human nature that is universal.

Westermarck's theory of incest was rejected by Sigmund Freud and others who believed that the incest taboo shows how moral rules arise as cultural inventions that suppress the immoral emotions of human nature.[56] Freud was a Hobbesian, in the sophistical tradition of ethical conventionalism, who saw human beings as so naturally selfish in their emotions that they could not live together in civilized societies unless they created cultural rules to subdue their natural inclinations. Freud thought that civilization required a rational rule of law to conquer "man's natural aggressive instinct, the hostility of each against all and of all against each."[57] Human beings must deny their animal nature through the moral imperatives of human culture as an autonomous realm of human rationality set apart from nature. The incest taboo was the most momentous manifestation of this human denial of nature, Freud insisted, because it was "the most drastic mutilation which man's erotic life has in all time experienced."[58] The incest taboo illustrates the general character of ethics as the rule of the "cultural super-ego" in demanding the renunciation of natural inclinations.[59] Ilham Dilman, in his book *Freud and Human Nature*, comments on the remarkable similarity of Freud's view of human nature to that taken by the sophist Callicles in Plato's *Gorgias*.[60]

Claude Lévi-Strauss spoke for the many social scientists who adopted this Freudian version of the sophistic position when he described the transcendent character of the incest taboo: "Before it, culture is still non-existent; with it, nature's sovereignty over man is ended. The prohibition of incest is where nature transcends itself."[61] By contrast, Westermarck believed that the incest taboo shows how moral rules arise as cultural practices that express human nature in manifesting the moral emotions. Like Plato, Aristotle, Aquinas, Smith, and Darwin, Westermarck saw human beings as naturally social animals with the natural emotions that fitted them for social life.

The debate between Westermarck and Freud over the origins of the incest taboo manifests a fundamental debate in the social sciences. For Westermarck, as representing the Socratic natural law tradition in social theory, human culture arises as the cultivation of human nature. For Freud, as representing the sophistic tradition in social theory, human culture arises as the conquest of human nature.

For Westermarck the incest taboo illustrates how social order arises from the complex interaction of nature and convention: the incest taboo is a social convention that expresses the human nature of the moral emotions. As a social convention, the incest taboo will vary across societies with diverse kinship systems. But as an expression of natural emotions, the incest taboo will show a natural propensity for most people to learn a sexual aversion for those with whom they have been reared from early childhood.

If Westermarck's ethical theory really is founded on an empirical science of human nature, as he suggests, then it should be subject to empirical confirmation or falsification. The debate over his theory of the incest taboo illustrates how his claims might be tested by scientific research. Beginning with Freud, Westermarck's theory has been dismissed with two criticisms. First, the occurrence of incest in all societies seems to indicate that there is no natural resistance to it. Second, if the taboo were natural, there would seem to be no need for cultural rules to enforce the taboo.[62] But these criticisms assume a simple dichotomy between fixed instinct and flexible culture that Westermarck denies. According to Westermarck, the instinctive propensity to incest avoidance is a tendency to learn sexual aversion when certain conditions are satisfied: most human beings are inclined to feel sexual aversion toward those with whom they have been reared from early infancy. Westermarck predicts, therefore, that most human beings raised in the familial environment typical for human beings will feel a strong aversion to incestuous relationships. But he also predicts that in some circumstances, some human beings will not acquire this aversion. For example, father-daughter incest is more likely to occur when the fathers have been separated from their daughters during their early rearing. Furthermore, he predicts that because of the natural variability in human emotional temperaments, a few human beings will not develop the aversion to incest that is normal for most people, and these deviant individuals will provoke a deep disgust from others. Because of this circumstantial and temperamental variability, human communities will develop cultural practices to enforce an incest taboo expressing the general feelings of the community in condemning those few who are inclined to commit incest.[63]

The incest taboo thus illustrates how social order arises from the complex interaction of nature and convention: the incest taboo is a social convention that expresses the human nature of the moral emotions. As a social convention, the incest taboo will vary across societies with diverse kinship systems. But as an expression of natural emotions, the incest taboo will show a natural propensity for most people to learn a sexual aversion for those with whom they have been reared from early childhood.

Arthur Wolf has indicated, in his survey of the scientific study of incest, that Westermarck's predictions seem to have been confirmed by the evidence. Wolf's special contribution to this research is his study of marriage in China. In parts of China, there were once three forms of marriage. In the "major" form of marriage, the bride went to live with her husband's family on the day of the wedding. In the "minor" form, a girl in infancy would join the family of her future husband as a *simpua*, or "little daughter-in-law," but she would not be married until she reached sexual maturity years later. In the "uxorilocal" form, the husband would submit to the authority of his father-in-law. From his meticulous study of marriage records in Taiwan, Wolf concluded that people in minor marriages showed far more sexual dissatisfaction than those in the other forms of marriage. They tended to produce more divorces, more adultery, and fewer children. He saw this observable behavior as showing that having been reared together in the same family from early infancy (age three or earlier), these spouses felt the

sort of sexual aversion to one another that would be predicted by Westermarck's hypothesis. Although they were not genetically related as brother
and sister, they displayed the same emotional discomfort with sexual union
that brothers and sisters typically feel toward one another. Natural selection has endowed us with a natural instinct to learn an emotional aversion
to sexual mating with those with whom we have been intimately associated
in our early years of rearing, because in the circumstances of evolutionary
history this would avoid the deleterious consequences of breeding with
close kin. This same natural propensity will produce such an aversion even
when the people with whom we have been reared are not our genetic kin.

The experience from the Israeli kibbutzim shows the same pattern.[64] In
kibbutzim with collective child-rearing, children from different families
grew up together from earliest infancy. Although not biologically related,
they lived with one another as if they were siblings. And although they
were permitted to marry, they never did, because they felt no sexual attraction to one another. As predicted by Westermarck, early childhood association inhibited sexual attraction.

Freud and other critics of Westermarck assumed that human beings were
the only animals that avoid incest, and thus it seemed that the incest taboo
must be a uniquely human cultural invention by which human beings subdue their animal emotions. Westermarck believed that many wild animals
are naturally inclined to avoid inbreeding.[65] Recent evidence from animal
behavior studies seems to support Westermarck. For example, monkeys and
apes show a tendency to avoid incest that is similar to that displayed by human beings.[66] As with human beings, incest does occur among other primates, but it is unusual, and it seems to arise only among exceptional individuals with abnormal temperaments or among those who have no close
bonding with their kin.[67] Among chimpanzees, the most common mechanism for avoiding inbreeding is that the females leave their natal group before breeding and join another group. (For most other primate species, the
males leave before breeding.) Some chimpanzee females, however, do not
leave their natal group; and all females live in the same group as their adult
sons. While there is some mating of immature individuals with immature or
adult relatives, mating of related adults is extremely rare. Mating is most
strongly inhibited between mothers and sons and between maternal siblings.
It seems, then, that chimpanzees have a natural propensity to avoid incest.[68]
Since chimpanzees are genetically closer to human beings than is any other
living species, it seems likely that incest avoidance among human beings
arises from a genetic propensity derived from a common ancestor.

Westermarck surveyed the biological research of his time suggesting that
inbreeding tended to produce high rates of infant mortality and of mental
and physical defects.[69] Modern genetic research confirms this conclusion.

Inbreeding increases the probability that deleterious recessive alleles in a population will be expressed, because any allele is more likely to be inherited simultaneously from both the paternal and maternal lines of a genealogy. This probability of producing genetically defective offspring increases in direct proportion to the closeness of the genetic relationship between two inbreeding individuals.[70]

But although inbreeding is costly, so is outbreeding. For example, traveling to a strange group to search for a mate might be costly. Consequently, some animals might be naturally disposed to seek a balance between inbreeding and outbreeding by choosing mates that differ somewhat but not too much from kin.[71]

To me all of this evidence provides convincing support for Westermarck's theory of the incest taboo. But it also seems to me that Westermarck's theory is compatible with the partial truth of the other major theories for explaining the incest taboo. Once the Westermarckian mechanism for learning incest avoidance is implanted in human nature by natural selection, then human beings might learn by experience that avoiding incest has many social benefits, and the recognition of those benefits might then provide utilitarian reasons for reinforcing the incest taboo. Incest avoidance mitigates sexual competition within the family. It promotes cooperation between kinship groups by encouraging intermarriage. And it avoids the harmful effects of inbreeding. There is some truth, therefore, to those theories of the incest taboo that emphasize these social benefits as likely motivations for human beings to deliberately enforce the taboo. But this is fully compatible with Westermarck's explanation for how the propensity to learn incest avoidance originally emerged in human evolutionary history. "Incest is considered harmful because it is disapproved of," Westermarck observes, "and it is not in the first place disapproved of because it is considered harmful."[72] This illustrates Adam Smith's claim that considerations of social utility influence our moral sentiments, although such utilitarian considerations cannot alone explain those sentiments.[73]

Even Freud's psychoanalytic theory might be partially true in a way that is compatible with Westermarck's theory, because if many of the patients that Freud saw had been reared by nurses in isolation from their parents, then Westermarck would predict that they would be inclined to develop incestuous urges that would need to be repressed. After all, the myth of Oedipus that so impressed Freud is the story of a man who had been separated from his parents at birth, who therefore did not experience the childhood familiarity with his mother that Westermarck believed was necessary to instill a sexual aversion to one's mother.[74]

The evidence for Westermarck's theory of the incest taboo comes from a variety of intellectual disciplines—sociology, anthropology, primatology, genetics, and evolutionary biology. In this way, it provides Edward Wilson with one of his best examples of what he calls "consilience"—the search for the unity of knowledge based on the idea that nature is governed by a seamless web of causal laws that cross the traditional disciplines of study, a search that began, Wilson thinks, with the ancient Greek philosophers.

Objections and Replies

Many scholars reject this idea of a comprehensive science of natural law that could explain human ethics as part of nature. In many cases, the objections to Wilson's Darwinian ethics are motivated by this fundamental rejection of natural law in ethics. For example, when Thomas Nagel argues that ethics as "an autonomous theoretical subject" cannot be explained by Wilson's sociobiology, he casually dismisses natural law theories because "they no longer have sophisticated adherents."[75] In fact, most of the objections to Darwinian ethics are restatements of the arguments John Stuart Mill used in his essay "Nature" against the idea of a natural moral law.[76] But it seems to me that the evidence and arguments for a Westermarckian theory of the incest taboo as rooted in the moral emotions of human nature suggest answers to the most common objections.

A survey of some of the recent critiques of Darwinian ethics—such as Paul Lawrence Farber's book *The Temptations of Evolutionary Ethics* and Peter Woolcock's essay "The Case Against Evolutionary Ethics Today"—would show that four objections are particularly prominent.[77] The most frequent objection to any ethical naturalism is that moral values cannot be derived from natural facts. This fact-value dichotomy is generally attributed to David Hume, who is said to have shown that there must be a radical separation between questions of what *is* or *is not* the case, which belong to the realm of nature, and questions of what *ought* or *ought not* to be done, which belong to the realm of morality. Because of this dichotomy, it seems to be a logical fallacy to infer a moral *ought* from a natural *is*. G. E. Moore called this "the naturalistic fallacy."[78] According to those who insist on such a dichotomy, natural science can describe the way things *are*, but it cannot prescribe the way things *ought* to be. A Darwinian science of human nature might describe the biological factors influencing human motivation, but it could not prescribe norms of proper human conduct without invoking moral standards that transcend the facts of human biology. For example, it would be proper for biologists to investigate the biological psychology of human sexuality, but for them to infer from the biological facts of human sexual motivation that some kinds of sexual conduct were morally better or worse than others would be fallacious. So, for instance,

biological scientists cannot—based on their scientific knowledge alone—
judge whether incest is right or wrong.

The second common objection to Darwinian ethical naturalism is that it
fails to recognize the primary role of reason in discovering objective principles for moral judgment. Darwinian ethics explains ethics as rooted in natural desires or emotions. But the critics argue that ethics requires the control of the desires or emotions by reason. The uniqueness of human ethics
seems to stem from the fact that human beings are the only animals who
can use reason to formulate and enforce objective rules of conduct that
transcend the subjective desires that govern the behavior of other animals.
In the attempt to ground ethics in animal emotions or desires, Darwinian
ethical theory cannot provide any rational foundation for ethics. Darwinian theorists explain human natural propensities to altruistic behavior, but
this does not explain the distinctly moral character of human social behavior as guided by rational principles of obligation.

The third objection is that the great variability in human social behavior
in different societies and in different historical periods denies the universal
human nature postulated by Darwinian ethics. The incest taboo seems to illustrate this very well. Just within the Anglo-American world, there is great
variation in the customary, religious, and legal rules of incest.[79] Until the six-
teenth century, the canon law of the Catholic Church as applied to England
prohibited marriage to a long list of both consanguineous and affinal relatives. Then, beginning in the Reformation, the list of prohibited relatives
was reduced drastically. In the twentieth century, the list was reduced even
further by eliminating most affinal relatives. In the United States, there has
been similar variation. For example, early in American history, marriage of
cousins was permitted (as it was in England). But beginning in the middle of
the nineteenth century, many states changed their laws to prohibit cousin
marriages. Now, however, the majority of the states allow cousin marriages.
If the incest taboo is governed by a biologically natural propensity, as Westermarck claims, then how could one explain such variation?

The fourth objection is that Darwinian reasoning about ethics assumes a
biological determinism that cannot account for the importance of human
culture and social learning in shaping ethical conduct in ways that transcend
biological nature. Ethical rules are not genetically determined in human beings. They arise from the habits and customs of human society. They are not
naturally innate but socially acquired. Consequently, the cultural diversity of
human beings creates a great diversity in moral rules that cannot be explained by reference to the biological universals of human nature.

In response to the first objection, I would accept Hume's "empiricist"
version of the fact-value dichotomy but not Kant's "transcendentalist" version. The Westermarckian ethical theorist would rightly agree with the

Humean understanding of the fact-value dichotomy if that means that moral judgments cannot be purely logical deductions from information about the world. But the Westermarckian theorist would disagree with the Kantian separation between the "phenomenal" realm of natural causality and the "noumenal" realm of moral freedom. The critics of Darwinian ethics often assume a Kantian notion of the "moral ought" or "the moral point of view" as utterly transcending the observable world of nature, but they rarely defend the metaphysical dualism implicit in this assumption, because it is hard to make such a radical dualism plausible.[80]

Hume distinguishes *is* and *ought* in order to show that moral assessments are derived not from pure reason alone but from moral emotions. Yet far from denying that moral judgments are judgments of fact, Hume rightly claims that moral judgments are accurate when they correctly report what our moral judgments would be in a given set of circumstances. Correct moral judgments are factual judgments about the species-typical pattern of moral sentiments in specified circumstances.

Hume compares moral judgments to judgments of secondary qualities such as color.[81] My judgment that this tomato is red is true if the object is so constituted as to induce the impression of red in normally sighted human beings viewing it under standard conditions. Similarly, my judgment that this person is morally praiseworthy is true if the person's conduct is such as to induce the sentiment of approbation in normal human beings under standard conditions. Just as an object can *appear* red to me when in fact it is not, so a person can *appear* praiseworthy to me when in fact he is not. The moral judgment whether some conduct would give to a normal spectator under standard conditions a moral sentiment of approbation is, Hume insists, "a plain matter of fact."[82] The moral sentiment itself, however, is a feeling or passion rooted in human nature that cannot be produced by reason alone. As Robert McShea has argued, moral judgments based on the facts of the moral emotions are not fallacious as long as they claim only that "for a particular intelligent species certain feelings are predictably aroused by certain facts and the experience of such feelings is the only basis on which we can make evaluative judgments."[83]

By contrast, Kant's version of the fact-value dichotomy separates reality into two metaphysical realms. Judging what *is* the case belongs to the "phenomenal" realm of nature, but judging what *ought* to be the case belongs to the "noumenal" realm of freedom. As moral agents, we obey categorical imperatives of what *ought* to be that are beyond the causal laws of nature.[84] "When we have the course of nature alone in view," Kant explains, "'*ought*' has no meaning whatsoever."[85]

Kant's separation of *is* and *ought* treats morality as an autonomous realm of human experience governed by its own internal logic with no reference to anything in human nature such as natural desires or interests. Contemporary philosophers follow Kant when they speak of "the moral point of view" as an autonomous realm of thinking that transcends factual reasoning about human nature. Kant does this because he accepts the sophistic and Hobbesian view of human nature. Since human beings are by nature selfish, asocial animals, they cannot live together in peace unless they conquer their natural inclinations by willing submission to moral rules devised by reason to pacify their selfish conflicts.[86]

This contrast between Humean empiricism and Kantian transcendentalism arises in the debate between Westermarck and Freud over the incest taboo. If the evidence supports Westermarck's theory of the incest taboo as opposed to Freud's, that would suggest that a moral rule like the incest taboo can be rooted in the natural propensities of human nature, and this would also suggest that the logical distinction between natural facts and moral values is not an absolute separation. Logical reasoning about the facts of sexual mating cannot by itself dictate a value judgment that incest is wrong, and that is what Hume means when he distinguishes *is* and *ought*. But if it belongs to the factual nature of most human beings that they have a natural propensity to acquire a feeling that incest is wrong, and if the generalization of that feeling across society supports a social prohibition of incest, then the value judgment that incest is wrong is indeed rooted in the facts of human nature.

The appeal to human nature in condemning social practices that violate the biological propensities for learning incest avoidance is evident in Arthur Wolf's defense of Westermarck's theory. Wolf says that the Chinese practice of "minor" marriages inflicted "wounds" on the people forced to marry against their will, because it was "contrary to their natural inclinations." This social practice was "abnormal," "maladaptive," "painful," "cruel," "oppressive," and "unnatural." Wolf insists: "It will be hard to find a more striking example of the fact that while culture can overwhelm natural tendencies, it is not capable of protecting the individuals involved from the suffering entailed."[87] Wolf acknowledges that a few individuals do not show the "Westermarck effect": some people who have been raised together in the same family from infancy develop no inhibition to sexual relations with one another. But he explains these exceptional individuals as being similar to psychopaths who suffer from an abnormal poverty of emotions so that they do not feel, or do not feel very deeply, the moral emotions typical for most human beings.[88] When Wolf thus appeals to the "natural inclinations" of emotionally "normal" people as setting standards for judging social practices as good or bad, he is implicitly appealing to the kind of natural law reasoning developed by Aquinas and others in the Socratic tradition of ethical naturalism.

When Wolf pronounces his moral condemnation of the practice of "minor" marriages in China, he assumes that most of his readers will agree with him. His masterful survey of the evidence and arguments for Westermarck's theory of incest avoidance offers inductive and deductive reasoning that will help to win his readers' agreement. But he cannot persuade them to accept his moral judgment purely through logical inference. He must assume that most of his readers have the typical emotional profile that will incline them to feel sympathy for the suffering of people forced into marriages made unhappy by their natural sexual aversion. Pure reason by itself could not elicit a moral judgment against "minor" marriages if such moral emotions were not brought into play. Readers who lack such emotions would not be moved to accept Wolf's moral judgment. Thus the moral appeal of Wolf's book—as combining logical reasoning with moral emotions —illustrates Westermarck's general claim about the primacy of natural emotions in moral experience.

This also suggests the proper response to the second objection to Darwinian ethical naturalism concerning the importance of reason in morality. As I indicated earlier, Westermarck—like others in the Socratic tradition— recognized the importance of reason in eliciting, directing, and generalizing the moral emotions. But he rightly denied that reason by itself could motivate ethical judgment. The incest taboo illustrates this very well. As long as the natural emotional aversion to incest creates the feeling that incest is wrong, reason can elicit, direct, and generalize that feeling in ways that will determine the customary and legal expression of the incest taboo. For example, human reason will work out elaborate systems of kinship classifications that will determine the details of whom one can and cannot marry. But if there were no original feeling in most human beings that incest is wrong, reason by itself by purely logical means could not judge that incest is wrong.

As indicated by the third objection, the variability in ethical rules such as the incest taboo is interpreted by many critics of Darwinian ethics as evidence that there are no universal ethical propensities rooted in human biological nature. But in fact, Westermarck's Darwinian account of the incest taboo illustrates how a Darwinian ethics can allow for both variability and universality.

Westermarck suggests that the incest taboo can expand beyond the nuclear family to embrace a wider circle of prohibited marriage partners or contract to cover a smaller circle, and this expansion or contraction will depend on the circumstances of familial living arrangements and kinship systems.[89] Originally, among the earliest human ancestors, Westermarck infers, the family consisted of parents and their children. The harmful consequences of inbreeding among such close relatives were such that natural selection would have shaped the sexual instinct so that people would tend to feel a sexual aversion toward those close relatives with whom they had lived from childhood. But once this natural propensity was acquired, it could be expressed as a sexual aversion toward remote relatives or even unrelated people who lived together in familial groups. Finally, this natural propensity could then lead "through an association of ideas and feelings" to the prohibition of sexual intercourse and marriage between people defined as kin who did not live together at all. Westermarck thus recognizes that the incest taboo varies greatly across societies and across time because of the variation in family life and kinship classification. But the variation shows an underlying regularity that manifests a natural biological tendency. For despite the variation in the incest rules for those outside the nuclear family of parents and children, the rule prohibiting sexual mating between parent and child and between siblings is almost universal. The rare cases of where the marriage of parents and children or of siblings is permitted are unusual exceptions that only highlight the universal rule. Any survey of the anthropological evidence will show that societies generally prohibit marriage between those people who typically live in the same household, which typically includes parents and children.

As rightly indicated by the fourth objection to Darwinian ethics, human reason gives us a flexibility in our capacity for symbolism and social learning that cannot be strictly determined by our genetic nature. But just as the Greek sophists invoked a false antinomy of nature versus convention, so do critics of Darwinian ethics such as Stephen Jay Gould insist on a false antinomy—biological potentiality versus biological determinism—that ignores the importance of biological propensity. Gould explains that his criticism of Edward O. Wilson "does not invoke a nonbiological 'environmentalism': it merely pits the concept of biological potentiality—a brain capable of the full range of human behaviors and rigidly predisposed toward none—against the idea of biological determinism—specific genes for specific behavior."[90] Explaining human social behavior through biological determinism is utterly implausible if that means claiming that specific genes determine specific behavior with no flexibility. Therefore, Gould asserts, we must accept the only alternative idea—"biological potentiality—a brain capable of the full range of human behaviors and rigidly predisposed toward none." But like the sophistical rhetoricians, Gould uses the word "rigidly" here to obscure a third alternative: the idea of biological predisposition or propensity as something more than a mere potentiality and yet something less than a rigid determinism.

With respect to sexual mating, for example, human beings have a biological potentiality for a wide range of behaviors—including celibacy, promiscuity, monogamy, polygyny, and polyandry. Gould would have a strong argument in claiming that there are no specific genes that absolutely determine or "rigidly" predispose us to one of these behaviors. But he would be wrong to assume from this that we have an indifferent potentiality for any of these. Although we have a potential for choosing complete celibacy, most human beings find this too difficult because it denies our strong propensity or desire for sexual mating. Promiscuity is easier because it caters to our sexual propensities. Polyandrous marriage seems to be a very weak potentiality for human beings because the intense sexual jealousy of males inclines them against it except in unusual circumstances. In contrast to polyandry, monogamous mating has been universal to all human societies, and polygynous mating has been common, because they satisfy biological desires. An understanding of biological propensities can explain why celibacy is difficult, promiscuity is easy, polyandry is rare, monogamy is universal, and polygyny is common, although none of these behaviors is "rigidly" determined by specific genes. As Wilson would say, this pattern of human behavior manifests the "epigenetic rules" of gene-culture coevolution. Our nature predisposes us to favor some mating behavior over others, although the specific expression of our mating behavior will reflect the variable conditions of physical environment, social circumstances, and individual temperament.

Similarly, the incest taboo manifests a natural propensity of human biological nature, although this taboo is not "rigidly" determined by specific genes. Most human beings are naturally inclined to learn a sexual aversion to those with whom they have lived in early childhood. This natural aversion will then tend to favor a universal taboo against incestuous relations between parents and children or between siblings. But the learning of this aversion will vary greatly in response to early childhood experiences. It will also vary in that a few individuals will not learn this aversion as easily as most others. Moreover, as we have seen, the extension of the taboo beyond the nuclear family is highly variable in response to varying kinship systems. Westermarck and Wilson emphasize this variability, and therefore they cannot properly be accused of a rigid biological determinism. But they also see a regularity in the incest taboo that reflects natural propensities. To talk as Gould does of an indifferent "potentiality" would acknowledge the cultural diversity in the incest taboo, but without explaining the underlying regularity in the taboo that Westermarck and Wilson explain as a propensity of human nature.

Conclusion

Westermarck's theory of the incest taboo explains that moral rule as a social convention that expresses the emotional propensities of human nature as shaped by natural selection in evolutionary history. That theory illustrates how a Darwinian science of human nature can support a naturalistic understanding of ethics as governed by natural causal laws. This provides a modern biological foundation for a tradition of ethical naturalism that begins with the Socratic philosophers of ancient Greece and extends to Thomas Aquinas in the Middle Ages and then to David Hume and Adam Smith in the early modern period. E. O. Wilson's view of ethics as rooted in the natural moral sentiments continues that tradition.

The fundamental alternative to such ethical naturalism is the ethical conventionalism of the sophistic tradition that stretches from Callicles to Hobbes and then to Freud and finally to contemporary theorists like Gould. According to this tradition, ethics requires a transcendence of the natural world in which we enter an autonomous realm of ethical conventions created by human reason. According to this view, there can be no natural science of ethics.

The scientific evidence and arguments for Westermarck's theory of the incest taboo show, by contrast, how the study of ethics could become an empirical science. This also suggests the possibility of a grand unification of all scientific knowledge in which ethics would become part of a science of human nature embedded within a general science of nature as a whole. A Darwinian science of ethics as natural right would thus allow us to understand our human nature as part of the natural cosmos.

NOTES

1. Edward Westermarck, *Ethical Relativity* (London: Kegan Paul, Trench, Trubner & Company, 1932), pp. 246–50, 288–89.

2. Edward O. Wilson, *Sociobiology: The New Synthesis* (Cambridge: Harvard University Press, 1975), p. 1.

3. Edward O. Wilson, *Consilience: The Unity of Knowledge* (New York: Alfred A. Knopf, 1998).

4. See Ullica Segerstrale, *Defenders of the Truth: The Battle for Science in the Sociobiology Debate and Beyond* (Oxford: Oxford University Press, 2000).

5. Wilson, *Consilience*, pp. 238–56.

6. Edward Wilson, *On Human Nature* (Cambridge: Harvard University Press, 1978), pp. 37–39, 68–69, 229; Charles J. Lumsden and Edward O. Wilson, *Genes, Mind, and Culture: The Coevolutionary Process* (Cambridge: Harvard University Press, 1981), pp. 37, 71, 85–86, 147–58, 238, 357; Charles J. Lumsden and Edward O. Wilson, *Promethean Fire: Reflections on the Origin of Mind* (Cambridge: Harvard University Press, 1983), pp. 64–65, 115, 119, 124–27, 133–38, 175–80; Michael Ruse and Edward O. Wilson, "Moral Philosophy as Applied Science," *Philosophy*, vol. 61 (Summer 1986): pp. 173–92; Wilson, *Consilience*, pp. 173–80.

7. Wilson, *Consilience*, p. 164.

8. Ibid., pp. 238–39.

9. Ibid., pp. 248–49.

10. Thomas Nagel, "Ethics as an Autonomous Theoretical Subject," in Gunther S. Stent, ed., *Morality as a Biological Phenomenon* (Berkeley: University of California Press, 1978), pp. 196–205.

11. David Buss, *The Evolution of Desire: Strategies of Human Mating* (New York: Basic Books, 1994), pp. 16–17.

12. Steven Pinker, *How the Mind Works* (New York: Norton, 1997), pp. 559, 561.

13. See Segerstrale, *Defenders of the Truth*, pp. 362–64.

14. Larry Arnhart, *Aristotle on Political Reasoning: A Commentary on the "Rhetoric"* (DeKalb: Northern Illinois University Press, 1981), pp. 118–34.

15. See Larry Arnhart, *Darwinian Natural Right: The Biological Ethics of Human Nature* (Albany: State University of New York Press, 1998).

16. Aristotle, *Sophistical Refutations*, 173a.

17. The sophistic tradition in ancient Greece is complex. For a good survey of its intellectual complexity, see G. B. Kerferd, *The Sophistic Movement* (Cambridge: Cambridge University Press, 1981). For the argument that modern ideas of cultural relativism and social constructionism continue the sophistic tradition, see Mark Backman, *Sophistication: Rhetoric and the Rise of Self-Consciousness* (Woodbridge, Conn.: Ox Bow Press, 1991).

18. Plato, *Laws*, trans. Thomas Pangle (New York: Basic Books, 1980), 838a–39b.

19. Xenophon, *Memorabilia*, IV.iv.19–23.

20. Stephen Salkever, *Finding the Mean: Theory and Practice in Aristotelian Political Philosophy* (Princeton, N.J.: Princeton University Press, 1990), p. 115.

21. Aristotle, *Nicomachean Ethics*, 1134b18–35a5; Thomas Aquinas, *Commentary on Aristotle's "Nicomachean Ethics"*, lec. 12.

22. Wilson, *Consilience*, p. 239.

23. Thomas Aquinas, *Summa Theologica*, I–II, q. 94, a. 2.

24. Aquinas, *Summa Theologica*, Suppl., q. 41, a. 1; q. 65, a. 1.

25. Aquinas, *Summa Contra Gentiles*, bk. 3, chap. 123.

26. Aquinas, *Summa Contra Gentiles*, bk. 3, chaps. 122–23.

27. Aquinas, *Summa Theologica*, II–II, q. 154, a. 9; Suppl., q. 54, a. 10; *Summa Contra Gentiles*, bk. 3, chap. 125.

28. William Durham, *Coevolution: Genes, Culture, and Human Diversity* (Stanford, Calif.: Stanford University Press, 1991), p. 331.

29. Aristotle, *History of Animals*, 630b31–31a7.

30. For surveys of the theories to explain the incest taboo, see Sybil Wolfram, *In-Laws and Outlaws: Kinship and Marriage in England* (London: Croom Helm, 1987), pp. 161–85; Carol R. Ember and Melvin Ember, *Anthropology*, 7th ed. (Englewood Cliffs, N.J.: Prentice-Hall, 1993), pp. 316–19; Durham, *Coevolution*, pp. 286–360; and Arthur Wolf, *Sexual Attraction and Childhood Association: A Chinese Brief for Edward Westermarck* (Stanford, Calif.: Stanford University Press, 1995), pp. 1–19, 498–515.

31. Thomas Hobbes, *The Leviathan*, chaps. 14–15.

32. Thomas Hobbes, *De Cive*, chap. 1.

33. Hobbes, *Leviathan*, chap. 17; Hobbes, *De Cive*, chap. 5, par. 5; Hobbes, *De Homine*, chap. 10.

34. On Hobbes's dualism, see Leo Strauss, *The Political Philosophy of Hobbes*, trans. Elsa M. Sinclair (Chicago: University of Chicago Press, 1952), pp. 7–9, 168–70.

35. Immanuel Kant, "Speculative Beginning of Human History," in *"Perpetual Peace" and Other Essays*, trans. Ted Humphrey (Indianapolis: Hackett Publishing Company, 1983); Immanuel Kant, *Critique of Judgment*, trans. Werner S. Pluhar (Indianapolis: Hackett Publishing Company, 1987), secs. 83–84.

36. See Adam Kuper, *Culture: The Anthropologists' Account* (Cambridge: Harvard University Press, 1999).

37. See Robert A. Greene, "Instinct of Nature: Natural Law, Synderesis, and the Moral Sense," *Journal of the History of Ideas* 58 (April 1997): pp. 173–94.

38. Bernard Mandeville, *The Fable of the Bees*, 2 vols. (Oxford: Oxford University Press, 1924), vol. 1: pp. 330–31.

39. Francis Hutcheson, *An Inquiry into the Original of Our Ideas of Beauty and Virtue* (London: J. Darby, 1725), p. 192.

40. Adam Smith, *The Theory of Moral Sentiments*, ed. D. D. Raphael and A. L. Macfie (Oxford: Oxford University Press, 1976), pp. 300–304, 321–27; Ian Simpson Ross, *The Life of Adam Smith* (Oxford: Oxford University Press, 1995), pp. 48–59, 159–64.

41. Smith, *Theory of Moral Sentiments*, pp. 163–66, 446–47.

42. See Wolf, *Sexual Attraction and Childhood Association*, pp. 463–75; and Mark T. Erickson, "Rethinking Oedipus: An Evolutionary Perspective of Incest Avoidance," *American Journal of Psychiatry*, vol. 150 (1993): pp. 411–16.

43. See Wolfram, *In-Laws and Outlaws*, pp. 30–40.

44. Charles Darwin, *The Descent of Man*, 2 vols. (London: John Murray, 1871), vol. 1: pp. 70–106, vol. 2: pp. 390–94.

45. Darwin, *Descent*, 1:70–71; Immanuel Kant, *The Critique of Practical Reason*, trans. Lewis White Beck (Indianapolis: Bobbs-Merrill, 1956), p. 90.

46. Charles Darwin, *The Origin of Species and The Descent of Man* (New York: Random House, Modern Library, 1936), pp. 485–86, 896.

47. Charles Darwin, *The Variation of Animals and Plants Under Domestication*, 2 vols. (New York: Appleton, 1897), vol. 2: p. 104.

48. Edward Westermarck, *The History of Human Marriage*, 3 vols., 5th ed. (New York: Allerton Book Company, 1922); Edward Westermarck, *The Origin and Development of the Moral Ideas*, 2 vols. (London: Macmillan Company, 1906); Westermarck, *Ethical Relativity*.

49. For an example of how a Westermarckian anthropological naturalism can support Thomistic natural law, see Thomas Davitt, *The Basic Values in Law* (Milwaukee, Wis.: Marquette University Press, 1978).

50. Westermarck, *Marriage*, vol. 1: pp. 22, 38, 53, 71, vol. 3: p. 365.

51. Westermarck, *Marriage*, vol. 3: pp. 104, 107, 206, 221–22; Westermarck, *Moral Ideas*, vol. 2: pp. 387–92.

52. For a study of Tibetan fraternal polyandry that largely confirms Westermarck, see Durham, *Coevolution*, pp. 42–102.

53. Westermarck, *Marriage*, vol. 2: pp. 82–239; Westermarck, *Moral Ideas*, vol. 2: pp. 364–81.

54. Westermarck, *Ethical Relativity*, pp. 248–49; Westermarck, *Marriage*, vol. 2: pp. 192, 197, 214.

55. For this formulation, see Wolf, *Sexual Attraction and Childhood Association*, p. 506.

56. Sigmund Freud, *Totem and Taboo*, in *The Standard Edition of the Complete Psychological Works of Sigmund Freud*, vol. 13, trans. James Strachey (London: Hogarth Press, 1955).

57. Sigmund Freud, *Civilization and Its Discontents*, trans. James Strachey (New York: Norton, 1961), pp. 42, 58, 62, 69, 87.

58. Ibid., p. 51.

59. Ibid., pp. 89–91.

60. Ilham Dilman, *Freud and Human Nature* (Oxford: Basil Blackwell, 1983).

61. Claude Lévi-Strauss, *The Elementary Structures of Kinship* (Boston: Beacon Press, 1969), pp. 30–31.

62. See Richard Lewontin, Steven Rose, and Leon Kamin, *Not in Our Genes: Biology, Ideology, and Human Nature* (New York: Pantheon, 1984), p. 137; and Philip Kitcher, *Vaulting Ambition: Sociobiology and the Quest for Human Nature* (Cambridge: MIT Press, 1985), pp. 280, 348.

63. Westermarck, *Marriage*, vol. 2: pp. 82, 88, 192, 201–3.

64. See Joseph Shepher, *Incest: A Biosocial View* (New York: Academic Press, 1983); and Wolf, *Sexual Attraction and Childhood Association*, pp. 435–38.

65. Westermarck, *Marriage*, vol. 2: pp. 223–24.

66. See Wolf, *Sexual Attraction and Childhood Association*, pp. 388–422; Anne Pusey, "Inbreeding Avoidance in Chimpanzees," *Animal Behavior*, vol. 28 (1980): pp. 543–52; Jane Goodall, *The Chimpanzees of Gombe* (Cambridge: Harvard University Press, 1986), pp. 466–71; Anne Pusey and Marisa Wolf, "Inbreeding Avoidance in Animals," *Trends in Ecology and Evolution*, vol. 11 (May 1996): pp. 201–6.

67. From her observations of wild chimpanzees, Jane Goodall reports, "Copulations between fathers and daughters and between paternal siblings are unlikely to be inhibited, for the individuals concerned do not 'know' their relationship. There is no close bonding between them, and they do not achieve the high level of familiarity that presumably underlies incest avoidance between mothers and sons and maternal siblings" (*Chimpanzees of Gombe*, 469).

68. See Anne Pusey, Chapter 3 in this volume.

69. Westermarck, *Marriage*, vol. 2: pp. 224–36.

70. See Durham, *Coevolution*, pp. 293–309.

71. See Patrick Bateson, "Optimal Outbreeding," in *Mate Choice*, ed. Patrick Bateson (Cambridge: Cambridge University Press, 1983), pp. 257–77.

72. Westermarck, *Marriage*, vol. 2: p. 182.

73. Smith, *Theory of Moral Sentiments*, pp. 179–93, 326.

74. The compatibility of Freud and Westermarck is argued by Robin Fox in *The Red Lamp of Incest* (Notre Dame, Ind.: University of Notre Dame Press, 1983).

75. Thomas Nagel, "Ethics as an Autonomous Theoretical Subject," in Gunther Stent, ed., *Morality as a Biological Phenomenon: The Presuppositions of Sociobiological Research* (Berkeley: University of California Press, 1980), p. 200.

76. John Stuart Mill, "Nature," in *Collected Works of John Stuart Mill*, ed. J. M. Robson (Toronto: University of Toronto Press, 1969), vol. 10: pp. 373–402.

77. Paul Lawrence Farber, *The Temptations of Evolutionary Ethics* (Berkeley: University of California Press, 1998); Peter G. Woolcock, "The Case Against Evolutionary Ethics Today," in Jane Maienschein and Michael Ruse, eds., *Biology and the Foundation of Ethics* (Cambridge: Cambridge University Press, 1999), pp. 276–306.

78. G. E. Moore, *Principia Ethica* (Cambridge: Cambridge University Press, 1903).

79. See Jack Goody, *The Development of the Family and Marriage in Europe* (Cambridge: Cambridge University Press, 1983); Wolfram, *In-Laws and Outlaws*; Carolyn S. Bratt, "Incest Statutes and the Fundamental Right to Marriage: Is Oedipus Free to Marry?" *Family Law Quarterly*, vol. 18 (1984): pp. 257–309; and Martin Ottenheimer, "Lewis Henry Morgan and the Prohibition of Cousin Marriage in the United States," *Journal of Family History*, vol. 15 (1990): pp. 325–34. For a vigorous argument for allowing cousin marriage in the United States, see Martin Ottenheimer, *Forbidden Relatives: The American Myth of Cousin Marriage* (Urbana: University of Illinois Press, 1996).

80. See, for example, Nagel, "Ethics as an Autonomous Theoretical Subject," and Woolcock, "Case Against Evolutionary Ethics Today."

81. David Hume, *A Treatise of Human Nature*, ed. L. A. Selby-Bigge (Oxford: Oxford University Press, 1888), p. 469; Hume, *Essays*, ed. Eugene F. Miller (Indianapolis: Liberty Classics, 1985), pp. 233–34.

82. David Hume, *Enquiries Concerning the Human Understanding and Concerning the Principles of Morals* (Oxford: Clarendon Press, 1902), p. 289.

83. Robert McShea, *Morality and Human Nature* (Philadelphia: Temple University Press, 1990), p. 226.

84. Immanuel Kant, *Critique of Practical Reason*, trans. Lewis White Beck (Indianapolis: Bobbs-Merrill, 1956), pp. 4–5, 18, 30–31, 99, 163–64; Kant, *Foundations of the Metaphysics of Morals*, trans. Lewis White Beck (Indianapolis: Bobbs-Merrill, 1959), pp. 4, 30, 44–45, 67–74, 80; Kant, *Critique of Pure Reason*, trans. Norman Kemp Smith (New York: St. Martin's Press, 1965), pp. 465, 472–79, 526; Kant, *Critique of Judgment*, pp. 286–87.

85. Kant, *Critique of Pure Reason*, p. 473.

86. Immanuel Kant, *"Perpetual Peace" and Other Essays*, pp. 87, 110–13, 115–17, 124–25; Kant, *Critique of Judgment*, pp. 317–21.

87. Wolf, *Sexual Attraction and Childhood Association*, pp. ix, 180, 217, 277–79, 328, 422.

88. See Wolf, *Sexual Attraction and Childhood Association*, pp. 259–63. For an account of psychopaths as people whose emotional poverty makes them "moral strangers," see Arnhart, *Darwinian Natural Right*, pp. 211–30.

89. Westermarck, *Marriage*, vol. 2: pp. 206–7, 214–18, 236–37.

90. Stephen Jay Gould, *Ever Since Darwin: Reflections in Natural History* (New York: Norton, 1977), pp. 257–58.

Contributors

LARRY ARNHART is Professor of Political Science at Northern Illinois University. He is the author of *Aristotle on Political Reasoning: A Commentary on the "Rhetoric"*; *Political Questions: Political Philosophy from Plato to Rawls*; and *Darwinian Natural Right: The Biological Ethics of Human Nature*. He is the associate editor of *The Encyclopedia of Science, Technology, and Ethics*.

PATRICK BATESON is Professor of Ethology at the University of Cambridge. He was Provost of King's College, Cambridge, for fifteen years and Biological Secretary of the Royal Society of London for five years. He edited, among other books, *Mate Choice* and is coauthor with Paul Martin of *Design for a Life: How Behavior and Personality Develop*.

ALAN H. BITTLES is Foundation Professor of Human Biology and Director of the Centre for Human Genetics at Edith Cowan University. During the last twenty-five years he has organized and conducted major studies into the prevalence and effects of consanguineous marriage, principally in South and Southeast Asia and North Africa. He has been editor of *Annals of Human Biology* since 1998.

WILLIAM H. DURHAM is Professor and Chair of Anthropological Sciences, Bing Professor in Human Biology, and Yang and Yamazaki University Fellow at Stanford University. He is author of *Scarcity and Survival in Central America* and *Coevolution: Genes, Culture and Human Diversity* and, since 1992, editor of the *Annual Review of Anthropology*.

MARK T. ERICKSON is Assistant Clinical Professor, University of Washington School of Medicine. He works and teaches at the Alaska Psychiatric Institute in Anchorage.

HILL GATES is Professor Emerita at Central Michigan University, has taught at Johns Hopkins and Stanford, and is now an independent scholar of Taiwan/China political economy and gender. Her principal publications include *Chinese Working Class Lives: Getting By in Taiwan*, *China's Motor: A Thousand Years of Petty Capitalism*, and *Looking for Chengdu: A Woman's Adventures in China*.

ANNE PUSEY is Professor of Ecology, Evolution, and Behavior and a Distinguished McKnight University Professor at the University of Minnesota. She was appointed Executive Director of the Jane Goodall Institute's Primate Research Programs in 2003 and has written over fifty scientific papers.

WALTER SCHEIDEL is Professor of Classics at Stanford University. He is author of *Measuring Sex, Age, and Death in the Roman Empire* and *Death on the Nile: Disease and the Demography of Roman Egypt*, editor of *Debating Roman Demography*, and coeditor of several volumes on the ancient economy and ancient empires.

NEVEN SESARDIC is Associate Professor in the Department of Philosophy at Lingnan University, Hong Kong. He has published papers in journals like *Philosophy of Science, British Journal for the Philosophy of Science, Biology, and Philosophy*, and *Ethics*. He is currently finishing a book, *Making Sense of Heritability*, which will be published by Cambridge University Press.

ARTHUR P. WOLF is Professor of Anthropological Sciences and David and Lucile Packard Foundation Professor in Human Biology at Stanford University. He is editor of *Religion and Ritual in Chinese Society* and *Family and Population in East Asia*; coauthor with Chieh-shan Huang of *Marriage and Adoption in Chinese Society, 1845–1945*; and author of *Sexual Attraction and Childhood Association: A Chinese Brief for Edward Westermarck*.

Index